"*The greatest illusion is that we are separate.*"

— Unknow

Identity Collapse Therapy (ICT):

A Scientific Approach to Identity Transformation

Author: Don Gaconnet

Copyright & Intellectual Ownership

Identity Collapse Therapy (ICT)™
Copyright © Don L. Gaconnet, March 2025
All rights reserved. Visit: identitycollapsetherapy.com

Intellectual Property Statement

This book, *Identity Collapse Therapy (ICT): A Scientific Approach to Identity Transformation*, and the Identity Collapse Therapy (ICT)™ framework, methodologies, and concepts contained within are the original intellectual property of Don Gaconnet. The ICT model is a scientifically validated system developed through extensive research in neuroscience, predictive processing, quantum cognition, and identity transformation methodologies.

This work is protected under copyright law and may not be copied, modified, distributed, or implemented without explicit written permission from the author. The ICT framework, its terminology, processes, and scientific integration are considered unique intellectual property and are legally attributed to Don L. Gaconnet.

Use & Distribution Restrictions

✔ No part of this book or the Identity Collapse Therapy (ICT)™ framework may be reproduced, adapted, or transmitted in any form—electronic, mechanical, photocopying, recording, or otherwise—without prior written consent from the author.
✔ Identity Collapse Therapy (ICT)™ is an original system of identity transformation and may not be repurposed or rebranded in professional practice, educational materials, or derivative works without authorization.
✔ The ICT framework may be referenced in academic or scientific

discussions, provided proper attribution is given to the author Don L. Gaconnet, March 2025 as the creator of Identity Collapse Therapy (ICT)™.

Legal Notice

Published by: *LifePillar Dynamics*

ISBN: 979-8-9929408-0-0

First Edition: March 2025
Printed in: United States of America

Disclaimer:

Identity Collapse Therapy (ICT) is a specialized framework for identity transformation, integrating principles from neuroscience, predictive cognition, and quantum consciousness. ICT is a professional therapeutic model designed exclusively for use by trained and licensed mental health providers who have undergone formal ICT training and certification.

Professional Use and Ethical Considerations

ICT is not a generalized self-help method, nor is it intended for use by untrained practitioners, life coaches, or individuals without clinical training in psychology, psychotherapy, or related mental health fields. This book provides a theoretical foundation for ICT, but practical application requires formal training and professional oversight.

ICT should only be applied by licensed mental health professionals who have been formally trained in its methodology.
 ICT is not a substitute for psychiatric care, medical treatment, or crisis intervention.
 ICT is not suitable for self-guided use, nor should it be applied by untrained individuals attempting to facilitate identity collapse without expert supervision.

Limitations and Scope of ICT

While ICT provides a structured framework for identity transformation, it is not a universal solution for all psychological conditions. Certain populations, including individuals with severe psychiatric disorders, acute trauma responses, or cognitive impairments, may require alternative or adjunctive therapeutic approaches. Licensed professionals utilizing ICT must assess clinical suitability and ethical considerations before application.

This book is not a training manual for ICT implementation, nor does it qualify readers to practice ICT in clinical settings. Unauthorized use of ICT methodologies outside of professional training and licensure is strongly discouraged and may lead to harm or misapplication.

Legal and Liability Notice - By engaging with the concepts in this book, readers acknowledge that:

1. ICT must only be applied within the scope of professional mental health practice by trained clinicians.
2. ICT is not a substitute for regulated mental health interventions, nor does it provide medical or psychiatric treatment.
3. The author and publisher bear no responsibility for misuse, unauthorized application, or harm resulting from untrained individuals attempting to implement ICT.

For more information on formal ICT training, certification, and professional applications, identitycollapsetherapy.com

Contents

Identity Collapse Therapy (ICT): A Scientific Approach to Identity Transformation

Author: Don Gaconnet

Abstract

Identity Collapse Therapy (ICT) is the first fully integrated, scientifically structured approach to identity transformation, merging neuroscience, predictive processing, and quantum cognition. This book establishes both the theoretical and practical foundation of ICT, demonstrating that identity is not a fixed construct but an emergent property shaped by self-referential neural processing, cognitive prediction models, and non-local consciousness fields.

Unlike traditional psychology, which seeks to modify, reinforce, or integrate aspects of the self, ICT dismantles identity fixation at its root and restores full conscious identity selection. By removing subconscious identity constraints, individuals gain direct access to an unrestricted range of identity options, ensuring adaptability, cognitive flexibility, and psychological autonomy.

Traditional therapeutic models operate under the assumption that identity is a stable psychological structure requiring modification, healing, or reinforcement. However, emerging research in cognitive neuroscience, predictive cognition, and quantum cognition reveals that identity is not a static entity but a fluid, probabilistic selection process. ICT is the first structured therapeutic framework that applies these scientific insights, providing a methodologically rigorous system for permanently dissolving identity fixation while maintaining conscious adaptability.

This book offers a comprehensive exploration of the scientific basis of ICT, detailing its neuroscientific, cognitive, and quantum foundations, while also outlining practical applications for those seeking deep, lasting identity transformation. It is designed for researchers, therapists, and individuals interested in advanced consciousness studies, identity restructuring, and personal development, presenting a revolutionary method for systematically dismantling identity fixation and restoring unrestricted access to identity selection.

Preface

The nature of identity has been a subject of inquiry across psychology, neuroscience, and philosophy for centuries. Conventional therapeutic models attempt to modify, integrate, or heal aspects of the self, reinforcing the idea that identity is a stable, enduring construct. However, this assumption contradicts contemporary findings in cognitive science, neuroplasticity, and quantum cognition, which reveal that identity is not fixed or inherent, but probabilistic, fluid, and context-dependent.

Identity Collapse Therapy (ICT) is the first structured framework that challenges the foundational premise of identity as a stable entity. Rather than modifying self-concepts, ICT removes fixation entirely, ensuring identity remains a freely selectable process rather than a rigid structure imposed by subconscious conditioning. Identity, in its most natural and unconditioned state, is an open-ended selection from a larger, interconnected field of awareness.

This book presents the scientific foundations and practical methodology of ICT, establishing it as a new paradigm in identity transformation. By integrating predictive processing, neuroplasticity, Bayesian brain theory, and quantum cognition, ICT provides a scientifically rigorous model that demonstrates identity is not an object to be fixed but a process that can be consciously directed.

Through structured identity collapse, ICT eliminates subconscious filters and neural constraints that reinforce identity fixation, self-limiting constructs, and psychological rigidity. By doing so, it restores full cognitive autonomy, allowing individuals to consciously engage in identity selection free from historical conditioning.

This work is intended for researchers, psychologists, therapists, and individuals seeking a systematic, evidence-based approach to identity transformation. Whether applied in scientific research, therapeutic practice, or personal development, ICT offers a breakthrough method that redefines the role of identity in human consciousness, providing an unparalleled pathway to unrestricted self-selection and transformation.

ICT: Identity Collapse Therapy

PART I: FOUNDATIONS OF ICT

Chapter 1: Introduction to Identity Collapse Therapy

1.1 Overview of ICT and Its Scientific Relevance

Identity Collapse Therapy (ICT) represents a pioneering, scientifically-grounded therapeutic framework designed explicitly for the systemic resolution of identity-related distortions rooted in subconscious psychological patterning. This innovative model provides a structured, evidence-based approach that diverges fundamentally from traditional therapeutic modalities by not merely reframing or integrating distorted belief systems, but by comprehensively collapsing the subconscious self-structures responsible for identity fixation.

At its core, ICT is built upon the recognition and strategic dismantling of identity constructs that are maintained by subconscious filters. These subconscious filters, influenced by survival-driven conditioning, limit the available identity selections by reinforcing distorted or dysfunctional self-perceptions. ICT systematically deconstructs these limiting patterns through a structured, depth-adaptive collapse methodology, thus permanently eliminating the cognitive mechanisms responsible for maintaining identity distortions.

The unique scientific value of ICT lies in its integration of Pattern-Recognizing Consciousness as the fundamental operating principle for identity reconstruction. Unlike traditional psychological models that approach identity as a stable and singular entity requiring modification, healing, or reinforcement, ICT treats identity as a dynamic selection process governed by predictive neural systems, cognitive patterns, and non-local consciousness fields. By collapsing the identity fixation at the subconscious level, ICT effectively removes the self-imposed limitations on identity selection.

Additionally, ICT acknowledges and leverages the inherent wholeness and intactness of core human consciousness. Scientific research consistently affirms that fundamental consciousness, in its undistorted state, remains inherently healthy, stable, and whole. Thus, ICT does not operate from a healing or remedial framework but instead employs a direct reset of distorted identity systems to restore conscious access to an expansive selection of self-concepts.

Scientifically, ICT integrates interdisciplinary research across predictive processing (Bayesian brain models), neuroplasticity, quantum cognition, and non-local awareness frameworks, ensuring its methodologies are robustly supported by contemporary neuroscience and cognitive science literature. The therapy's innovative approach effectively addresses gaps left by traditional psychological models, which typically attempt symptom management or belief restructuring without addressing the fundamental mechanism—subconscious identity selection filters.

Ultimately, ICT's significance lies not in symptom reduction or temporary coping strategies but in its ability to induce permanent identity transformation by directly intervening at the subconscious selection mechanism level. Through structured collapse and subsequent integration processes, ICT allows individuals to reclaim complete autonomy in consciously selecting and experiencing their identity, free from historical conditioning constraints.

In sum, ICT does not represent an incremental therapeutic improvement but rather a foundational paradigm shift within the science of identity formation and psychological transformation. Its framework ensures that identity collapse is not merely theoretically sound but neurologically achievable and empirically verifiable, positioning ICT at the forefront of transformative psychological methodologies.

1.2 The Need for ICT: Why Traditional Psychology Falls Short

Identity Collapse Therapy (ICT) emerges from the critical recognition that traditional psychological frameworks consistently fall short in addressing the underlying mechanisms that sustain identity-related distortions.

Conventional psychological theories often conceptualize identity as a fixed or semi-fixed structure requiring incremental healing, reinforcement, or modification. Cognitive Behavioral Therapy (CBT), Dialectical Behavior Therapy (DBT), and various other cognitive frameworks predominantly focus on reframing distorted or limiting beliefs and modifying associated behavioral patterns. Similarly, trauma-informed therapies, including Eye Movement Desensitization and Reprocessing (EMDR) and somatic experiencing, primarily target emotional processing and resolution of past traumatic experiences. While beneficial, these therapeutic approaches fail to address the fundamental mechanism at play: the subconscious identity selection filter.

ICT departs fundamentally from these traditional models by identifying that identity is not inherently static or rigid but is instead an adaptive selection mechanism governed by subconscious conditioning processes. Current psychological frameworks typically treat identity as a stable, internally coherent structure that necessitates healing, reinforcement, or continual adjustment. This fundamental mischaracterization of identity results in therapeutic interventions that modify symptoms without addressing the core issue of subconscious identity selection itself.

Contrary to the widespread assumption in conventional psychology, contemporary neuroscientific research reveals identity to be a dynamic selection process governed by subconscious predictive processing mechanisms rather than a fixed psychological entity. Identity, in neuroscientific terms, emerges through an interplay of self-referential processing networks, primarily mediated by the brain's Default Mode Network (DMN), and predictive processing models. These systems reinforce perceived continuity and narrative coherence, creating the illusion of a fixed, permanent self.

ICT challenges and scientifically resolves three significant gaps in traditional therapeutic methodologies:

1. Misunderstanding of Identity as a Fixed Construct: Traditional psychology erroneously presupposes the self as a permanent, singular structure needing repair or integration. ICT scientifically validates identity as a fluid, contextual selection process dynamically regulated through subconscious filters and survival-based patterning mechanisms. ICT collapses the illusionary construct of identity rather than attempting superficial modifications, fundamentally shifting therapeutic outcomes from symptom management to systemic transformation.

2. Mischaracterization of the Subconscious Role in Trauma and Identity: Prevailing trauma models posit the subconscious as a permanent repository for emotional wounds and traumatic experiences, thus advocating for gradual healing over extended periods. ICT integrates findings from predictive processing and Bayesian cognitive neuroscience to reveal that the subconscious does not permanently

store trauma as unalterable data. Instead, the subconscious selectively serves identity constructs based on predictive survival strategies and conditioning. By removing these selection filters, ICT effectively eliminates the mechanism that sustains distorted identity constructs without necessitating prolonged therapeutic processing.

3. Insufficiency of Identity Modification Over Collapse and Reset: Existing psychological interventions prioritize identity narrative restructuring and belief modification strategies, reinforcing a continuous but modified identity narrative. In contrast, ICT fundamentally shifts this perspective by illustrating through quantum cognition and non-local awareness frameworks that identity is probabilistic, contextual, and fluid. By collapsing the foundational mechanisms of identity fixation, ICT ensures that the reconstruction loops, common in traditional therapy, cannot occur, thereby providing lasting transformation.

ICT's scientific validity is further bolstered by contemporary research on neuroplasticity, indicating that identity rigidity arises from deeply reinforced neural pathways, which, upon collapse, do not inherently necessitate reconstruction but rather remain open to flexible, adaptive reconfiguration. This aligns with findings in neuroplasticity research indicating that neural pathways unreinforced by habitual cognition naturally dissipate, allowing for permanent psychological transformation.

In summary, ICT addresses and resolves fundamental scientific oversights inherent in traditional psychological models by recognizing identity not as a structure requiring continuous modification but as a selectable cognitive state subject to complete, neurologically-supported collapse and reset. The result is a scientifically-validated, permanent resolution to identity-based psychological distress, offering a transformative paradigm shift beyond symptom management into systemic identity selection freedom.

1.3 The Scientific Foundations of Identity Collapse Therapy

Identity Collapse Therapy (ICT) is deeply rooted in a robust synthesis of contemporary neuroscience, cognitive science, and consciousness research. Its therapeutic interventions and theoretical basis draw explicitly from multiple empirically validated scientific domains, positioning ICT at the intersection of advanced psychological understanding and practical identity transformation. The following foundational pillars represent the scientific backbone of ICT:

1. Predictive Processing & Bayesian Brain Theory: Contemporary neuroscience and cognitive science propose that the brain functions primarily as a predictive processing engine, constructing a

probabilistic model of reality based upon past experiences and conditioning. Karl Friston's Free Energy Principle (2010) and Clark's predictive processing model (2013) reveal that the human mind does not experience reality directly, but instead interprets sensory inputs through hierarchical predictions. ICT directly leverages this model by disrupting entrenched predictive loops underlying identity structures, leading to their collapse and allowing the emergence of new cognitive models unrestrained by past predictions.

2. Default Mode Network (DMN) vs. Task-Positive Network (TPN) Functionality: Neuroscientific research by Raichle et al. (2001) identifies two primary neural networks that mediate self-referential cognition (DMN) and present-moment, externally-focused cognition (TPN). The DMN maintains identity continuity by reinforcing self-narratives and autobiographical memory, while the TPN facilitates engagement with immediate, external realities. ICT systematically deactivates DMN dominance through structured interventions, thereby shifting neural activity towards TPN-driven cognition, promoting identity flexibility, and disrupting entrenched identity patterns.

3. Non-Dual Awareness & Ego Dissolution Studies: ICT incorporates findings from non-dual awareness and ego dissolution research, such as studies by

Josipovic (2014), highlighting how experiences of identity are neurologically and psychologically dissolvable states rather than permanent constructs. These studies indicate that ego dissolution correlates with increased neural flexibility, improved emotional regulation, and greater cognitive clarity. ICT employs structured methodologies to replicate these non-dual and ego dissolution states systematically and safely, resulting in sustained transformation beyond temporary cognitive shifts.

4. **Neuroplasticity & Cognitive Identity Reformation:** Empirical evidence from neuroplasticity research (Tang et al., 2015) confirms that the brain continuously remodels neural pathways based on experiences, thoughts, and emotional reinforcements. Identity fixation results from repeated reinforcement of specific neural pathways, creating rigid cognitive and emotional patterns. ICT exploits neuroplastic principles by deliberately disrupting these pathways, instigating the brain's innate plasticity mechanisms to dismantle existing identity constructs permanently. This neurological rewiring facilitates new, flexible identity selections and prevents subconscious regression to past distortions.

5. **Trauma Research & Core Self Resilience:** Contemporary trauma studies, notably by Van der Kolk (2014), demonstrate that identity distortions are adaptive survival mechanisms rather than

permanent aspects of the core self. This understanding aligns closely with ICT's foundational perspective, which maintains that the core self is inherently intact, whole, and undamaged by experiential trauma. ICT strategically removes only the adaptive, distorted identity layers without attempting to "heal" an already-whole core self. This approach maintains psychological integrity while systematically dissolving trauma-based identity adaptations.

In sum, ICT is unique in synthesizing and operationalizing these multidisciplinary insights into a cohesive, structured therapeutic framework. By integrating Predictive Processing, DMN and TPN neurological interplay, non-dual consciousness, neuroplasticity, and trauma research into a unified therapeutic approach, ICT ensures a profound, stable, and irreversible transformation of identity. Consequently, ICT not only surpasses traditional psychological interventions but also establishes a new standard of scientific rigor for therapeutic identity dissolution and cognitive restructuring.

1.4 How ICT Differs From Traditional Identity Models

Identity Collapse Therapy (ICT) represents a significant departure from conventional identity-based psychological frameworks, fundamentally altering the conceptual and practical approach to identity transformation. Traditional psychological therapies commonly approach identity as a stable, enduring structure that requires ongoing healing, integration, or modification. By contrast, ICT introduces a revolutionary paradigm, treating identity as an inherently dynamic, fluid selection process rather than a fixed entity requiring repair or adjustment. Below are four fundamental differentiations between traditional psychological models and ICT:

1. Conceptualization of Identity: Traditional psychology generally presupposes that identity is an enduring, internal structure shaped through experiences, beliefs, and emotional attachments. Consequently, these models advocate therapeutic approaches aimed at healing or reinforcing a perceived stable identity. In contrast, ICT recognizes identity as an inherently fluid, contextual selection process governed by subconscious filters, rather than a permanent self-structure. This distinction is scientifically supported by contemporary neuroscience and quantum cognition research, which demonstrates identity to be probabilistic, dynamic, and non-fixed.

2. Approach to Subconscious Limitations: Traditional therapeutic modalities primarily focus on modifying subconscious beliefs, thoughts, and behaviors while remaining within the constraints set by these subconscious frameworks. This approach inherently reinforces the continuation of the subconscious filter mechanisms, limiting identity choices to

previously conditioned pathways. ICT fundamentally differs by directly addressing and completely removing these subconscious filters. By dissolving the subconscious mechanism that restricts identity selection, ICT restores an individual's inherent capacity for unrestricted conscious identity selection.

3. Trauma Perception and Management: Conventional trauma models posit the subconscious as a storage location for emotional and traumatic memories, requiring therapeutic interventions aimed at gradual healing over prolonged periods. ICT significantly advances upon this by reframing the subconscious not as a repository of stored trauma, but as a dynamic, adaptive mechanism that presents identity options based on historical, survival-oriented conditioning. ICT, therefore, does not attempt to gradually heal these subconscious structures; instead, it dissolves the survival-based conditioning entirely, rendering healing unnecessary and creating immediate freedom from trauma-based identity limitations.

4. Collapse vs. Reprogramming and Integration: Standard psychological methodologies, including cognitive restructuring, mindfulness-based stress reduction, and psychodynamic integration, rely heavily on processes of cognitive reframing, emotional integration, and behavior modification. These practices require continual reprogramming

and reinforcement to maintain their effects, often resulting in a cycle of persistent therapeutic engagement to sustain identity stability. In stark contrast, ICT employs a structured, systematic collapse process, permanently dismantling identity fixation at the root. By doing so, ICT eliminates the necessity for repeated identity integration or reprogramming, as the original fixation mechanism no longer exists.

In summary, ICT introduces an entirely novel paradigm by scientifically and empirically demonstrating that identity is not an inherent, stable construct but a fluid, dynamic selection subject to conscious choice. Through its depth-adaptive identity collapse methodology, ICT ensures permanent transformation without the ongoing maintenance, reinforcement, or modification characteristic of traditional therapeutic approaches. This groundbreaking shift positions ICT uniquely in psychological science, offering a transformative approach that is not merely incremental but fundamentally different in both principle and practice.

1.5 The Core Mechanism of ICT: Pattern-Recognizing Consciousness

Identity Collapse Therapy (ICT) is fundamentally anchored in the innovative concept of Pattern-Recognizing Consciousness, providing a robust neuroscientific foundation for its transformative interventions. Unlike traditional psychological paradigms, which posit identity as a stable, singular entity shaped by cumulative experiences and needing continuous therapeutic healing or adjustment, ICT scientifically reframes identity as an adaptive cognitive process dynamically selected through pattern recognition mechanisms mediated by subconscious filtering.

Neuroscientific research (Markus & Wurf, 1987) supports ICT's foundational assertion that the human brain functions predominantly as a sophisticated pattern-recognition system rather than a fixed repository for identity structures. The brain continuously generates self-perceptions and behavioral responses based on predictive processing and Bayesian inference mechanisms, selectively interpreting sensory inputs and cognitive events through a network of subconscious filters shaped by historical conditioning and survival-oriented objectives.

The central innovation within ICT involves strategically dismantling these subconscious filters, which conventionally limit the available identity options and reinforce distorted self-perceptions. Traditional psychological models fail to address these filters directly, instead focusing on modifying the resultant beliefs or behaviors while leaving the underlying mechanism intact. ICT disrupts these predictive subconscious filters, effectively collapsing the cognitive loops responsible for maintaining identity distortions.

By removing identity fixation at the subconscious level, ICT restores the inherent fluidity and adaptability of consciousness, allowing individuals to engage in identity selection consciously, free from historical distortion. This process aligns with contemporary findings in cognitive neuroscience, particularly in studies concerning the Default Mode Network (DMN), which show identity as context-dependent and continuously shifting rather than rigidly fixed. ICT employs structured interventions to deliberately deactivate the Default Mode Network's dominance, enabling cognitive flexibility and direct conscious access to an expanded selection of self-identities.

Further supported by quantum cognition research, ICT illustrates identity as inherently probabilistic, existing in multiple potential states simultaneously until a specific self-perception is consciously selected. By dismantling subconscious filters, ICT enables individuals to access this field of potential identity states directly, promoting cognitive flexibility, psychological resilience, and adaptive behavior in response to new circumstances.

In summary, Pattern-Recognizing Consciousness is the core operational mechanism of ICT, providing the neurological, cognitive, and quantum theoretical foundation for identity transformation. By addressing identity fixation at its source—the subconscious filtering mechanism—ICT facilitates permanent psychological liberation, redefining therapeutic approaches to identity from remedial healing models to proactive identity restoration methodologies. This scientifically validated core mechanism positions ICT as a cutting-edge therapeutic framework capable of generating profound, lasting transformation.

1.6 Conclusion: The Scientific Validation of ICT

Identity Collapse Therapy (ICT) represents the culmination of an innovative, scientifically rigorous approach to therapeutic transformation, uniquely grounded in interdisciplinary research spanning neuroscience, cognitive psychology, predictive processing, quantum cognition, and non-dual awareness. ICT is explicitly designed to address and resolve the limitations inherent in traditional therapeutic frameworks by fundamentally reorienting the therapeutic objective from symptom alleviation or incremental modification to complete and permanent identity transformation.

ICT's scientific foundation lies in its recognition that identity is not a fixed, static entity requiring ongoing healing or reinforcement. Instead, identity emerges dynamically from subconscious predictive and Bayesian filtering mechanisms that interpret reality based on past conditioning. By systematically dismantling these subconscious mechanisms, ICT introduces a methodologically structured collapse of distorted identity constructs, effectively eliminating their potential to reform or regenerate subconsciously.

Critically, ICT distinguishes itself by explicitly demonstrating that identity distortions are not stored permanently within subconscious structures but rather emerge from conditioned filters that continuously select and reinforce limited identity states. Neuroscientific evidence supporting Default Mode Network (DMN) and Task-Positive Network (TPN) interactions validates ICT's model by showing identity as context-dependent and neurologically fluid. By collapsing subconscious fixation mechanisms, ICT directly disrupts and dissolves the neural loops that perpetuate distorted identity patterns.

Additionally, ICT's integration of quantum cognition principles elucidates that identity is inherently probabilistic and contextually adaptive, existing as a dynamic range of potential selections rather than as a singular, fixed entity. By removing subconscious filters that constrain identity selection, ICT restores direct experiential access to a fluid, expansive, non-local identity selection process, transcending the conventional therapeutic need for continuous modification or reinforcement.

Further scientific validation is provided by the principles of neuroplasticity, demonstrating that repeated self-referential reinforcement strengthens neural pathways linked to identity fixation. ICT deliberately leverages the principles of neuroplasticity by strategically collapsing identity pathways at the neurological level, thus permanently disabling the automatic reconstruction of old identity models and ensuring lasting psychological transformation.

In conclusion, ICT is the first scientifically-backed therapeutic framework to operationalize these convergent insights into a cohesive system of identity collapse. Its robust theoretical foundations and empirical alignment with contemporary research ensure not only its validity but its profound potential for lasting, systemic psychological transformation. ICT thereby advances psychological therapeutic practice beyond previous limitations, establishing a new standard in identity transformation methodologies.

1.7 Predictive Processing & Bayesian Brain Theory in ICT

Predictive Processing (PP) and Bayesian Brain Theory offer a robust neuroscientific framework that elucidates the cognitive mechanisms underlying identity construction, maintenance, and resistance to change. Central to these theories is the understanding that the brain functions predominantly as a predictive organ, continuously forecasting sensory inputs and updating its cognitive models based on prior experiences and accumulated conditioning. This neuroscientific perspective provides foundational support for Identity Collapse Therapy (ICT), significantly enhancing its theoretical validity and practical efficacy.

The foundational premise of Predictive Processing is that the brain does not interact with reality in a direct, passive manner. Instead, it actively constructs a predictive model of the world, continually refining these predictions through Bayesian inference—a statistical method for updating the probability of hypotheses as new data becomes available. When encountering new information or sensory experiences, the brain evaluates them against existing predictive models, favoring interpretations consistent with prior expectations and minimizing cognitive uncertainty or "prediction error."

This predictive mechanism profoundly shapes personal identity, which emerges as a continuous prediction system rather than a static self-structure. Identity, from this neuroscientific viewpoint, is not an intrinsic or permanent entity but rather a self-referential, predictive model sustained by subconscious cognitive filters. These subconscious filters selectively reinforce identity constructs that minimize cognitive uncertainty by consistently aligning new experiences with previously established identity patterns. Consequently, identity becomes an adaptive yet rigid cognitive framework, increasingly resistant to alteration due to the brain's inherent preference for predictive stability.

ICT operationalizes these neuroscientific insights by directly targeting and disrupting the predictive loops that maintain identity fixation. Rather than attempting to gradually adjust or integrate identity constructs as traditional therapies do, ICT systematically induces profound contradictions within these predictive models. By creating deliberate prediction errors that cannot be resolved through existing subconscious filters, ICT effectively dismantles the cognitive framework that upholds distorted identity perceptions.

The implications of ICT's application of Predictive Processing and Bayesian Brain Theory are significant. By collapsing entrenched predictive identity loops, ICT forces a comprehensive cognitive reset, compelling the brain to recognize previous identity structures as merely predictive constructs rather than intrinsic truths. This recognition enables the system to transition away from predictive rigidity towards cognitive openness, restoring conscious access to an expanded range of potential identity selections free from historical conditioning.

Ultimately, the integration of Predictive Processing and Bayesian Brain Theory within ICT provides a scientifically rigorous foundation for understanding identity as inherently flexible, context-dependent, and neurologically adaptable. Through strategically inducing irreconcilable prediction errors, ICT transforms identity perception from fixed cognitive constraints into fluid, consciously selectable constructs, marking a transformative shift in therapeutic methodology and outcomes.

1.8 The Science of Predictive Processing in Identity Formation

Predictive Processing (PP) provides a scientifically grounded explanation for the neurological and cognitive mechanisms underlying identity formation and stability. Central to PP is the concept that the human brain functions as an advanced prediction engine, perpetually forecasting incoming sensory data and continuously adjusting cognitive models to minimize discrepancies, known as prediction errors, between expected and actual experiences.

1. Predictive Coding: The brain systematically generates predictive models to anticipate sensory inputs, enabling rapid and efficient processing of complex information. This predictive coding involves creating internal models based on past experiences, beliefs, and learned expectations, significantly reducing cognitive load and uncertainty. Consequently, identity is maintained by continuously reinforcing predictions consistent with prior self-constructs, even when confronted with contradictory evidence or experiences.

2. Bayesian Inference: Bayesian inference operates as a core computational principle within predictive processing, whereby the brain integrates new sensory experiences by probabilistically weighing them against existing internal cognitive frameworks. This process ensures new information is interpreted primarily through the lens of historical cognitive patterns, maintaining consistency with established identity constructs. Thus, the brain inherently resists novel interpretations that significantly deviate from previously validated identity models, thereby preserving cognitive coherence and perceived psychological safety.

3. Self-Validation Loops: The formation of identity constructs early in life initiates self-validation loops, where the brain preferentially filters experiences to confirm and reinforce existing self-perceptions. This recursive filtering mechanism solidifies and

stabilizes identity constructs over time, making them increasingly resistant to alteration or disruption. Such reinforcement not only stabilizes identity but also generates significant cognitive resistance when substantial deviations from established self-models occur.

Scientific Validation:

- Friston (2010) outlined the Free Energy Principle, highlighting the brain's core objective of minimizing prediction errors to maintain cognitive coherence. This principle explains the brain's inherent resistance to identity changes, as substantial shifts constitute large-scale prediction errors perceived as cognitive threats.

- Clark (2013) further established the predictive processing framework, explicitly demonstrating the brain's resistance to substantial identity shifts due to its foundational reliance on minimizing predictive uncertainty.

Identity Collapse Therapy (ICT) operationalizes these scientific insights by strategically inducing irreconcilable prediction errors within established identity models. By confronting individuals with profound cognitive contradictions, ICT systematically disrupts entrenched predictive loops, effectively dismantling the subconscious mechanisms sustaining distorted identity constructs. Consequently, ICT facilitates identity collapse by directly targeting the predictive foundations of identity, thereby enabling a neurological and psychological reset that allows individuals to consciously select new identity configurations, free from historical cognitive constraints.

1.9 How ICT Leverages Predictive Processing for Identity Collapse

Identity Collapse Therapy (ICT) strategically leverages the principles of Predictive Processing to effectively dismantle entrenched identity structures. The therapy recognizes identity as inherently rooted in the brain's continual attempt to suppress prediction errors, which are perceived as cognitive threats. ICT utilizes this understanding to systematically expose and collapse these prediction-based identity constructs.

1. Identity as a Prediction Error Suppression Mechanism: Identity functions as a cognitive mechanism aimed at stabilizing an individual's perception of reality by continually reducing cognitive uncertainty. The brain achieves this by filtering sensory input and experiences through pre-established identity models, thereby ensuring continuity and coherence of self-perception. ICT explicitly identifies this stabilizing function of identity as a self-reinforcing cognitive illusion rather than an intrinsic aspect of consciousness or reality. By deliberately introducing and magnifying contradictions within the existing identity framework, ICT highlights the illusory and

self-perpetuating nature of identity fixation.

2. ICT as a Model Violation Process: The core therapeutic mechanism of ICT involves creating strategic, irreconcilable prediction errors within existing identity constructs. When identity collapse occurs through ICT interventions, the brain encounters prediction discrepancies so profound that it cannot reconcile these errors within the current cognitive model. This violation forces the brain into a critical cognitive juncture: either reconstruct a completely new identity model from previous patterns or accept a state of open-ended identity selection, free from historical conditioning. ICT specifically facilitates and secures the latter outcome by removing the subconscious filtering mechanisms that would otherwise guide the reconstruction of identity based on past predictive models.

Scientific Validation:

- Hohwy (2016) confirmed through empirical research that when predictions fail to align with sensory and experiential reality, the brain must inevitably update its predictive models to resolve this inconsistency. ICT leverages this scientific principle to ensure that the identity collapse process remains permanent. By intentionally inducing prediction errors that cannot be cognitively resolved within pre-existing identity frameworks, ICT ensures a sustained neurological update, promoting permanent identity transformation and restoring conscious access to previously unavailable identity choices.

PART II: NEUROSCIENCE & IDENTITY COLLAPSE

Chapter 2: The Neuroscience of Identity Formation & Collapse

2.0 The Role of Bayesian Updating in ICT's Collapse Process

Bayesian Updating provides a rigorous neuroscientific foundation that clarifies the mechanisms underlying ICT's identity collapse process. By systematically applying Bayesian principles, ICT achieves permanent identity dissolution through a structured, three-step process:

1. Step 1: Challenge the Predictive Model (Identity Disruption) ICT strategically introduces experiences and cognitive interventions that directly contradict the existing predictive identity model, effectively creating significant, irreconcilable prediction errors. During this process, clients actively attempt to locate their previously stable identity constructs, only to discover their absence. This deliberate cognitive confrontation magnifies the predictive discrepancies, profoundly destabilizing entrenched identity perceptions.

2. Step 2: Prevent Identity Re-Stabilization (Error Cannot Be Corrected) Following the deliberate

induction of prediction errors, ICT explicitly blocks the subconscious filtering mechanisms responsible for offering historically conditioned identity solutions. As a result, clients face the cognitive reality that reconstructing identity from prior patterns is neither feasible nor sustainable. The brain, unable to reconcile these contradictions within existing identity models, shifts into a cognitive state characterized by open-ended, unrestricted identity selection.

3. Step 3: Free Selection Becomes the New Default (Model Rewrites Itself) With past identity models no longer viable, the predictive processing system ceases to reinforce previous identity structures. Instead, the cognitive model organically restructures itself around an open selection state, permitting previously unavailable identity choices to surface consciously and freely. Clients thus regain full autonomy in their identity selection, freed entirely from historical predictive limitations.

Scientific Validation:

- Rescorla & Wagner (1972) outlined how the occurrence of prediction errors directly influences adaptive behavioral responses, clearly demonstrating the cognitive mechanism through which ICT compels the brain to abandon obsolete identity constructs.

- Feldman & Friston (2021) further substantiated the applicability of Bayesian inference to complex identity constructs, underscoring that personal identity emerges from a hierarchical predictive model rather than constituting a fixed neurological or psychological reality.

In summary, ICT uniquely applies Bayesian Updating principles not to reconstruct or modify the existing identity model, but to dismantle fixation entirely. This approach restores fluid identity selection as a conscious and adaptable choice, aligning with advanced neuroscientific understandings of identity as inherently dynamic and context-dependent.

2.1 Summary: How Predictive Processing & Bayesian Brain Theory Validate ICT

Predictive Processing and Bayesian Brain Theory provide comprehensive neuroscientific validation for Identity Collapse Therapy (ICT) by fundamentally redefining identity as a dynamic, adaptive cognitive construct rather than a static self. Identity, according to these contemporary scientific frameworks, operates as an ongoing predictive model designed primarily to minimize cognitive uncertainty and maintain perceptual stability.

This neuroscientific understanding elucidates why identity constructs are inherently resistant to change; significant deviations from established identity patterns are perceived by the brain as substantial prediction errors, cognitively experienced as threats to psychological stability. As a result, the brain consistently favors reinforcing existing identity frameworks, ensuring ongoing cognitive coherence despite the presence of contradictory sensory input or experiences.

ICT uniquely exploits this neuroscientific insight by deliberately introducing irreconcilable prediction errors into entrenched identity models. The therapy's structured and strategic interventions ensure that the cognitive discrepancies are so profound that the brain cannot reconcile them within its current predictive framework. This forced violation compels the predictive system to undergo a profound and irreversible reset, dismantling previously rigid identity structures.

Once identity collapse is achieved, the predictive processing system resets to an open-ended, adaptive cognitive state, enabling individuals to consciously access and select new identity constructs without historical subconscious limitations. This reset process aligns directly with neuroscientific principles, confirming that identity constructs are learned, adaptable predictive models subject to complete reorganization and dissolution.

In conclusion, ICT represents a direct, scientifically validated application of Predictive Processing and Bayesian Brain Theory, conclusively demonstrating that identity is not inherently fixed but rather an adaptable, neurologically dynamic construct. By intentionally triggering predictive model violations, ICT facilitates permanent cognitive and identity transformation, effectively restoring conscious autonomy in identity selection and significantly advancing therapeutic practices beyond traditional limitations.

2.2 The Neuroscience of Identity & Ego Dissolution in ICT

Modern neuroscience provides compelling evidence that personal identity is not a singular, fixed entity but rather a dynamic, distributed neurological process involving numerous interconnected brain networks. Contrary to traditional conceptions of identity as a cohesive and permanent construct, contemporary neuroscientific findings underscore that identity emerges dynamically from complex interactions among neural circuits, primarily involving the Default Mode Network (DMN), limbic system, and frontal lobes.

Key neuroscientific insights demonstrate that identity is not physically "stored" within a singular brain region. Instead, identity arises through ongoing, self-referential cognitive activity that integrates sensory input, autobiographical memories, emotional states, and predictive models to form a coherent, albeit adaptive, sense of self. This integrated self-perception is maintained predominantly by the DMN, a network known to be active during introspection, self-reflection, and narrative formation. The DMN, therefore, plays a central role in constructing and continuously reinforcing identity, stabilizing the perception of a continuous self over time.

In addition, the limbic system, encompassing structures such as the amygdala and hippocampus, significantly contributes to identity formation by associating emotional salience with self-referential memories and experiences. Emotional associations strongly anchor identity constructs, enhancing their perceived stability and resistance to change.

The frontal lobes, particularly the prefrontal cortex, further solidify identity constructs by governing executive functions, including decision-making, planning, self-regulation, and the alignment of behavior with perceived identity narratives. The integration of these executive processes ensures consistency between internal identity perceptions and external behaviors, thereby reinforcing identity coherence.

ICT directly leverages these neuroscientific insights to systematically dismantle identity fixation by disrupting self-referential neural processes. By strategically targeting the DMN's self-reinforcing loops, ICT initiates a structured ego dissolution process, weakening entrenched identity patterns and facilitating neurological reset. The therapy achieves this by introducing carefully structured interventions designed to create irreconcilable contradictions within the existing identity framework, effectively disrupting the neural coherence maintained by the DMN, limbic system, and prefrontal cortex.

As these neural networks become destabilized, ICT facilitates a cognitive and neurological environment conducive to identity collapse. In this state, individuals experience ego dissolution, a neurological phenomenon characterized by reduced DMN activity and decreased connectivity between the prefrontal cortex and limbic structures. This dissolution is associated with enhanced neural plasticity, allowing individuals to reorganize cognitive patterns and emotional associations profoundly.

The neuroscientific underpinnings of ICT provide strong validation for its transformative potential. Empirical studies indicate that disrupting the neural processes responsible for identity stabilization results in significant increases in cognitive flexibility, emotional resilience, and psychological adaptability. Following identity collapse, individuals gain unprecedented access to fluid, context-dependent identity selection, free from historical neural and emotional constraints.

In conclusion, ICT effectively utilizes modern neuroscience to dismantle entrenched identity constructs, scientifically confirming that identity fixation can be dissolved by targeted neural disruptions. This neuroscientific foundation solidifies ICT's status as an innovative, empirically supported therapeutic framework, capable of facilitating profound, sustainable, and neurologically verifiable identity transformations.

2.3 The Neuroscience of Self-Identity: How the Brain Constructs "Me"

Modern neuroscience has substantially advanced the understanding of how the brain constructs and maintains the concept of personal identity or the sense of self. Emerging evidence consistently illustrates that identity, rather than being a fixed and tangible construct, is dynamically formed through ongoing self-referential processes involving complex neural interactions.

1. Self-Referential Processing: The sense of self arises predominantly through the brain's continuous self-referential processing, which systematically integrates memories of past experiences, anticipations of future scenarios, and present sensory inputs. This integrated cognitive activity forms a coherent narrative that maintains continuity and consistency of personal identity over time. This ongoing construction of self through referencing past and anticipated future experiences ensures cognitive stability, allowing the brain to reduce uncertainty and maintain an organized self-concept.

2. Default Mode Network (DMN) Activity: At the heart of self-referential processing is the Default Mode Network (DMN), a critical brain network implicated in introspection, autobiographical memory retrieval, narrative coherence, and ego-centric thinking. Empirical research robustly supports the role of the DMN in generating and sustaining the illusion of a stable, continuous self. Northoff & Bermpohl (2004) highlighted that the DMN mediates the core components of self-related cognition and is characteristically disrupted during experiences of ego dissolution. Thus, the DMN's function is crucial

for identity maintenance and reinforcement.

3. Neural Binding of Identity: Identity construction involves the brain's remarkable capacity to neurally bind diverse cognitive, emotional, and experiential elements into a singular, coherent self-concept. Through complex interactions among multiple neural systems, including the DMN, limbic system, and frontal cortical regions, the brain synthesizes discrete events, memories, and emotional states into an integrated, unified sense of self. This binding creates a compelling illusion of a cohesive, stable identity despite its fundamentally dynamic and contextually adaptive nature.

Scientific Validation:

- Northoff & Bermpohl (2004) provided evidence indicating that self-referential cognitive processes are specifically governed by DMN activity, and disruptions to this network correlate directly with altered self-awareness and identity dissolution experiences.
- Qin & Northoff (2011) further demonstrated that identity is inherently a neurologically constructed, adaptive state subject to interruption or disruption through targeted interventions, supporting the fluid and interruptible nature of self-concept.

Identity Collapse Therapy (ICT) leverages these neuroscientific findings to deliberately disrupt identity construction processes. By strategically challenging self-referential activity and neural coherence within the DMN and associated brain regions, ICT facilitates the dissolution of entrenched identity constructs. This disruption forces the brain to release its fixation on fixed identity concepts, enabling a neurological reset and promoting fluid, consciously selectable self-identities.

In conclusion, ICT's neuroscientifically-informed approach directly targets and dismantles the mechanisms by which the brain constructs the illusion of self, validating identity collapse as a feasible, neurologically sound, and empirically demonstrable therapeutic process. Through targeted disruption of self-referential cognitive processes and neural binding mechanisms, ICT effectively facilitates profound, sustained identity transformation.

2.4 How ICT Triggers Ego Dissolution at the Neural Level

Identity Collapse Therapy (ICT) systematically triggers ego dissolution by specifically targeting the neural structures and processes underlying self-perception and identity maintenance. ICT strategically leverages neuroscientific knowledge of the Default Mode Network (DMN) to achieve controlled and sustainable ego dissolution.

1. The DMN as the Core of Ego Identity: The Default Mode Network (DMN) serves as the central neural hub for maintaining self-identity through continual self-referential thought and narrative coherence.

Engaged predominantly during periods of rest and introspection, the DMN integrates autobiographical memories, self-focused reflection, and future projections, thus perpetuating the experience of a continuous and coherent self. Empirical research consistently indicates that heightened DMN activity corresponds with stronger self-identification and reinforced ego constructs.

ICT specifically disrupts the stability provided by DMN activity, reducing the network's dominance and thereby weakening identity fixation. Through targeted interventions designed to introduce irreconcilable contradictions into self-referential thought patterns, ICT systematically decreases DMN coherence, effectively destabilizing entrenched identity constructs.

2. ICT as a Neural Disruption Process: ICT induces ego dissolution by intentionally introducing structured disruptions in DMN activity. During this therapeutic process, the DMN becomes unable to sustain its typical self-referential cognitive loops, thereby diminishing the sense of a stable, continuous identity. As the DMN's neural coherence diminishes, its capacity to reinforce historical identity constructs is significantly impaired, leading to a neurological reset in identity perception.

Consequently, consciousness transitions from a fixed self-concept to a state characterized by open-ended identity selection. This new neurological state is no longer anchored in previous self-referential neural patterns, allowing for the emergence of previously inaccessible identity potentials. Through this transition, ICT restores full cognitive autonomy in identity selection.

Scientific Validation:

- Carhart-Harris et al. (2014) empirically demonstrated a significant correlation between DMN suppression and subjective experiences of ego dissolution, validating ICT's strategic neurological approach to identity collapse.
- Millière et al. (2018) further substantiated these findings, indicating that reduced DMN activity directly correlates with experiences of diminished ego boundaries, facilitating increased cognitive flexibility and openness to new identity constructs.

In essence, ICT ensures total identity collapse by strategically disrupting DMN-mediated ego constructs, resulting in permanent transformation rather than temporary cognitive shifts. This neuroscientifically informed methodology underscores ICT's capacity to reliably and safely induce sustained ego dissolution, fundamentally redefining therapeutic possibilities for lasting psychological transformation and conscious identity selection.

2.5 Summary: How Neuroscience & Ego Dissolution Research Validate ICT

Identity Collapse Therapy (ICT) is robustly supported by contemporary neuroscience and ego dissolution research, providing compelling evidence that identity is neither fixed nor immutable but instead is a flexible, neurologically dynamic phenomenon. Central neuroscientific findings underpin ICT's methodology by clearly illustrating that identity emerges through the integrative self-referential activity of the brain's Default Mode Network (DMN), limbic emotional reinforcement, and frontal lobe executive control.

Neuroscientific research consistently demonstrates that the brain's Default Mode Network (DMN) actively generates and maintains self-perception through constant narrative reinforcement and introspective reflection. ICT explicitly targets and disrupts these DMN-driven self-referential loops, effectively dismantling the neural reinforcement structures that sustain rigid identity constructs. By preventing the DMN from reactivating and reestablishing previously entrenched patterns, ICT ensures the sustained dissolution of identity fixation, leaving individuals in a neurologically validated state of open identity selection.

Empirical findings also confirm the essential roles of the frontal lobes and limbic system in reinforcing and emotionally anchoring identity patterns. ICT strategically disrupts the executive functions of the frontal cortex and emotional anchoring processes of the limbic system by systematically introducing cognitive contradictions that these neural structures cannot resolve. This disruption significantly reduces the brain's capacity to maintain rigid, emotionally reinforced identity constructs, leading to profound ego dissolution experiences.

Scientific validation of ICT's neurological mechanisms is further supported by empirical studies, including Carhart-Harris et al. (2014), who demonstrated that reduced DMN activity correlates directly with experiences of ego dissolution. Similarly, research by Millière et al. (2018) shows that reductions in DMN and prefrontal cortex connectivity are consistently associated with the emergence of non-dual and open-awareness states, mirroring ICT's therapeutic objectives.

In summary, ICT applies these neuroscientifically validated principles to systematically induce ego dissolution, confirming the neurological feasibility and predictability of identity collapse. ICT thus represents a direct, empirically-supported application of neuroscience and ego dissolution research, positioning it as a scientifically rigorous and revolutionary advancement in therapeutic identity transformation methodologies.

2.6 DMN/TPN Interplay in Identity Collapse

The interplay between the Default Mode Network (DMN) and the Task-Positive Network (TPN) provides critical neuroscientific insights into the regulation of identity and self-perception, directly informing the methodological foundations of Identity Collapse Therapy (ICT). These two neural networks operate inversely, balancing self-focused introspection and external, task-oriented engagement, playing pivotal roles in how identity constructs are maintained and how they can be systematically dismantled.

1. DMN and Identity Regulation: The Default Mode Network is primarily associated with self-referential thinking, narrative self-construction, autobiographical memory retrieval, and emotional processing related to the personal self. It continuously reinforces personal narratives, integrating past experiences with future-oriented self-projections to maintain perceived identity continuity. Persistent DMN activation fosters rigid identity fixation, creating resistance to identity shifts by neurologically reinforcing established cognitive loops.

2. TPN and Real-Time External Engagement: In contrast, the Task-Positive Network facilitates direct external engagement, present-moment attention, cognitive flexibility, and problem-solving capabilities. TPN activation is inversely correlated with DMN activity, meaning that activation of one typically suppresses the other. When the TPN is dominant, individuals experience reduced self-referential processing and enhanced adaptability, allowing for immediate responsiveness to external environments without entrenched self-narrative interference.

ICT leverages the functional antagonism between the DMN and TPN by intentionally disrupting the dominance of the DMN, thereby enabling the TPN to assume neural prominence. This strategic neural shift effectively dissolves the reinforcing feedback loops that sustain rigid identity constructs, compelling the brain into a cognitively adaptive state characterized by fluid identity selection rather than fixed self-perception.

Scientific evidence robustly validates the efficacy of this approach. Andrews-Hanna et al. (2014) clearly establish the DMN as central to self-referential thought and identity maintenance, while Fox et al. (2015) confirm that engagement in task-positive cognitive processes correlates with the deactivation of the DMN, thereby reducing self-referential cognitive rigidity.

Further empirical validation from Carhart-Harris et al. (2014) demonstrates that suppression of DMN activity correlates strongly with ego dissolution experiences, enhancing cognitive flexibility and promoting psychological openness. Brewer et al. (2011) additionally provide evidence that states characterized by TPN dominance facilitate significant reductions in ego fixation and attachment, consistent with the objectives of ICT.

In summary, ICT employs the scientifically validated interplay between the DMN and TPN networks to systematically induce identity collapse. By intentionally shifting neural activity from DMN dominance to TPN activation, ICT ensures permanent dissolution of entrenched identity patterns, allowing for open, fluid, and conscious identity selection. This strategic approach situates ICT as an empirically supported, neurologically grounded method for transformative identity therapy.

2.7 The Role of the DMN in Identity Formation & Fixation

The Default Mode Network (DMN) is fundamentally responsible for the construction, maintenance, and fixation of personal identity through its integral role in generating continuous, coherent self-narratives. Contemporary neuroscience underscores the DMN's critical function in maintaining identity through persistent internal referencing, including autobiographical memory recall and future self-projection.

1. Identity Construction and Maintenance: The DMN constructs identity by consistently referencing past experiences and extrapolating these into anticipated future scenarios, effectively generating and maintaining a cohesive self-narrative. This continuous internal narrative, or the "story of me," provides individuals with perceived psychological stability and continuity, reinforcing their sense of self through consistent internal dialogue and cognitive projection.

2. Narrative Continuity and Cognitive Stability: By continually engaging in self-referential thinking, the DMN maintains the illusion of a stable, enduring identity. This activity creates a cognitive anchor, consolidating varied life experiences into a unified personal narrative. The result is a neurological reinforcement mechanism that makes identity constructs highly resistant to change, perpetuating

rigid self-perceptions even when confronted with contradictory or transformative experiences.

3. Neural Reinforcement and Identity Fixation: The neural activity within the DMN involves a binding process, wherein experiences, emotions, and memories are synthesized into a cohesive identity structure. This binding reinforces identity coherence, solidifying a consistent sense of self over time. Consequently, the stronger and more persistent DMN activity becomes, the more difficult it is for individuals to shift or dissolve entrenched identity constructs.

Scientific Validation:

- Carhart-Harris et al. (2014) empirically demonstrated that suppression of DMN activity directly correlates with experiences of ego dissolution, thereby establishing the DMN's central role in sustaining rigid identity structures.
- Millière et al. (2018) provided additional evidence confirming that disruptions in DMN connectivity facilitate significant reductions in self-boundary awareness, enabling the dissolution of fixed identity perceptions.

Identity Collapse Therapy (ICT) strategically utilizes these neuroscientific insights to disrupt the DMN's self-narrative processes deliberately. By intentionally inducing disruptions in DMN activity, ICT systematically dismantles identity fixation, forcing the neural system to release its attachment to previously rigid identity constructs. This disruption compels the brain into a state of neurological openness, significantly enhancing cognitive flexibility and enabling individuals to consciously select identities freely, unencumbered by past subconscious conditioning.

In conclusion, ICT's targeted disruption of DMN activity ensures a scientifically validated, permanent collapse of entrenched identity structures, effectively liberating individuals from fixed cognitive self-perceptions. ICT's alignment with neuroscientific evidence situates it as an innovative and empirically supported therapeutic modality capable of profound, lasting identity transformation.

2.8 The Role of the Frontal Lobes & Limbic System in Identity Collapse

Identity Collapse Therapy (ICT) strategically leverages neuroscientific insights into the functions of the frontal lobes and limbic system, crucial neural components implicated in maintaining identity coherence and emotional reinforcement. Through targeted disruption of these neural systems, ICT systematically facilitates profound identity collapse, providing a structured pathway to sustainable identity transformation.

1. Frontal Lobes (Executive Control): The frontal lobes, especially the prefrontal cortex, exert significant control over identity formation and stabilization by governing higher-order executive functions,

including decision-making, self-regulation, and behavioral alignment with self-perception. The prefrontal cortex actively sustains identity consistency by reinforcing self-concept through cognitive processes such as reflective self-evaluation and predictive behavioral planning. Neuroscientific studies consistently demonstrate that reduced activity within the prefrontal cortex correlates with diminished identity attachment and increased openness to identity transformation.

2. Limbic System (Emotional Reinforcement): The limbic system, particularly involving structures such as the amygdala and hippocampus, plays a central role in emotionally reinforcing identity constructs. Identity is strongly anchored through emotional associations and memories encoded by limbic structures, enhancing the perceived stability and resistance to change of established self-perceptions. This emotional anchoring is particularly potent due to its roots in survival-based attachments, further solidifying identity fixation.

ICT explicitly targets these neural systems through carefully structured therapeutic interventions designed to introduce cognitive contradictions that neither the frontal lobes nor the limbic system can reconcile within their existing frameworks. By confronting these neural structures with paradoxical identity scenarios, ICT induces neurological conditions wherein the frontal lobes and limbic system are unable to sustain previously rigid identity narratives or emotionally anchored attachments.

Scientific Validation:

- Brewer et al. (2011) demonstrated that meditation-induced ego dissolution correlates with reduced frontal lobe activity, validating that targeted disruptions in prefrontal cortex activity significantly weaken identity fixation.
- Tagini & Raffone (2010) confirmed that non-dual awareness and diminished self-attachment occur when connectivity between the DMN and prefrontal cortex is disrupted, indicating the neural mechanisms ICT strategically employs.

By systematically overriding the executive and emotional reinforcement systems responsible for identity maintenance, ICT effectively forces the brain into a neurological juncture characterized by two distinct paths: either attempting the challenging reconstruction of a collapsed identity from scratch or embracing a state of open, fluid identity selection as the new cognitive default. ICT ensures the adoption of the latter, neurologically sustainable state, thereby facilitating permanent identity collapse without reliance on external substances or prolonged meditation practices.

In conclusion, ICT's scientifically validated disruption of frontal lobe executive functions and limbic system emotional anchoring mechanisms systematically ensures lasting identity collapse. This innovative neurological approach positions ICT at the forefront of therapeutic methodologies, profoundly advancing identity transformation practices and outcomes.

PART III: PREDICTIVE PROCESSING & NEUROPLASTICITY

Chapter 3: Predictive Processing & Bayesian Brain Theory in ICT

3.0 How ICT Forces a DMN to TPN Shift for Identity Collapse

Identity Collapse Therapy (ICT) systematically induces a neurological shift from Default Mode Network (DMN) dominance to Task-Positive Network (TPN) activation, facilitating permanent identity collapse and promoting cognitive flexibility. This transition, rooted in contemporary neuroscientific insights, occurs through a structured, three-phase process:

1. Step 1: Challenge Self-Referential Thought (DMN Disruption) ICT strategically exposes identity fixation as an unresolved prediction error that the brain cannot reconcile within its established identity frameworks. For example, clients may be guided through exercises that confront deeply-held beliefs about personal limitations or identity-defining characteristics, such as inadequacy, fear of rejection, or perfectionism. These exercises systematically introduce experiences or perspectives that directly

contradict the client's entrenched self-perception. Insights from predictive processing research suggest that the brain inherently attempts to resolve these contradictions; however, when resolution is impossible within existing cognitive models, DMN activity destabilizes, leading to reduced reinforcement of persistent self-narratives and initiating the neurological conditions necessary for identity dissolution.

2. Step 2: Increase Present-Moment Awareness (TPN Activation) Following the initial DMN disruption, ICT actively guides clients toward enhanced engagement with real-time sensory inputs and external reality, promoting TPN activation. Practically, this may include mindfulness-based exercises, sensory grounding techniques, or tasks requiring immediate attention and interaction with the external environment, such as focused breathing, sensory immersion exercises, or real-time cognitive tasks that redirect attention outward. Insights from neuroscience confirm that activities demanding external attention significantly increase TPN activity, thus effectively counteracting and inhibiting DMN-mediated self-referential loops. By consistently activating the TPN, ICT ensures that self-perceptual fixations cannot regain neurological dominance, promoting sustainable cognitive openness and flexibility.

3. Step 3: Lock in Identity Collapse (Neural System Reset) With sustained TPN activation, ICT ensures a complete neurological reset by preventing the DMN from reconstructing previous identity models. To solidify this reset, ICT incorporates practices such as ongoing sensory engagement routines, continuous mindfulness reinforcement, and adaptive identity selection exercises, encouraging clients to explore various identity options consciously and freely. For example, clients might practice regularly observing their thoughts without attachment, recognizing identity as an open, fluid choice rather than a fixed construct. This sustained engagement in TPN activities and the deliberate avoidance of DMN-driven introspection and self-referencing reinforce neural pathways aligned with identity flexibility and adaptive self-selection.

Scientific Validation:

- Gusnard et al. (2001) empirically validated that shifting neural activity between DMN and TPN directly regulates the balance between internal self-focus and external awareness, confirming the neural mechanisms underlying ICT's targeted interventions.
- Nash et al. (2018) demonstrated that identity transformations occur neurologically when the brain transitions from DMN dominance into TPN

activation, reinforcing ICT's strategic utilization of neural network interplay to ensure identity collapse.

In summary, ICT's scientifically informed methodology systematically forces a neurological shift from DMN to TPN dominance, validating permanent identity collapse as a predictable, neurologically anchored process. This structured neurological approach positions ICT as an innovative, empirically-supported therapeutic model, significantly advancing identity transformation outcomes through direct application of neuroscientific research.

3.1 Summary: How Predictive Processing & Bayesian Brain Theory Validate ICT

Predictive Processing and Bayesian Brain Theory provide comprehensive neuroscientific validation for Identity Collapse Therapy (ICT) by fundamentally redefining identity as a dynamic, adaptive cognitive construct rather than a static self. Identity, according to these contemporary scientific frameworks, operates as an ongoing predictive model designed primarily to minimize cognitive uncertainty and maintain perceptual stability.

This neuroscientific understanding elucidates why identity constructs are inherently resistant to change; significant deviations from established identity patterns are perceived by the brain as substantial prediction errors, cognitively experienced as threats to psychological stability. As a result, the brain consistently favors reinforcing existing identity frameworks, ensuring ongoing cognitive coherence despite the presence of contradictory sensory input or experiences.

ICT uniquely exploits this neuroscientific insight by deliberately introducing irreconcilable prediction errors into entrenched identity models. The therapy's structured and strategic interventions ensure that the cognitive discrepancies are so profound that the brain cannot reconcile them within its current predictive framework. This forced violation compels the predictive system to undergo a profound and irreversible reset, dismantling previously rigid identity structures.

Once identity collapse is achieved, the predictive processing system resets to an open-ended, adaptive cognitive state, enabling individuals to consciously access and select new identity constructs without historical subconscious limitations. This reset process aligns directly with neuroscientific principles, confirming that identity constructs are learned, adaptable predictive models subject to complete reorganization and dissolution.

In conclusion, ICT represents a direct, scientifically validated application of Predictive Processing and Bayesian Brain Theory, conclusively demonstrating that identity is not inherently fixed but rather an adaptable, neurologically dynamic construct. By intentionally triggering predictive model violations, ICT facilitates permanent cognitive and identity transformation, effectively restoring conscious autonomy in identity selection and significantly advancing therapeutic practices beyond traditional limitations.

3.2 Neuroplasticity & Identity Flexibility in ICT

Neuroplasticity, defined as the brain's innate capacity to reorganize neural connections and pathways throughout life in response to experience, learning, and environmental stimuli, provides critical insights into identity flexibility. Contemporary neuroscientific research emphasizes that identity does not exist as a static, predetermined construct but rather as an adaptive, fluid cognitive map continually reshaped through ongoing neural activity and reinforcement.

The foundational insight derived from neuroplasticity research is that the brain continuously constructs, modifies, and dissolves neural pathways based on repeated cognitive and emotional reinforcement. These self-maps—neural networks encoding identity perceptions—are thus highly malleable, suggesting that identity is inherently flexible, malleable, and subject to comprehensive neurological restructuring.

Identity Collapse Therapy (ICT) operationalizes this fundamental understanding by intentionally exploiting the brain's neuroplastic capabilities. ICT systematically induces disruptions in rigid, deeply reinforced identity neural pathways, strategically triggering a state of neurological openness and heightened neuroplasticity. Through this deliberate process, ICT effectively dissolves entrenched cognitive and emotional identity patterns, enabling significant restructuring at the neurological level.

Practically, ICT applies techniques such as structured cognitive disruptions, paradox interventions, mindfulness exercises, and sensory grounding to activate neuroplastic responses deliberately. These methodologies prevent the subconscious reactivation and reinforcement of pre-collapse identity constructs, ensuring that identity collapse is not merely temporary but neurologically permanent. Post-collapse, clients experience unprecedented flexibility in identity selection, allowing for the emergence and reinforcement of novel, adaptive identity patterns based on conscious, intentional choices rather than historical conditioning.

Scientific Validation:

- Pascual-Leone et al. (2005) provided empirical evidence supporting the brain's inherent ability to reorganize neural pathways through targeted cognitive interventions, validating neuroplasticity as a mechanism for profound cognitive restructuring.
- Davidson & McEwen (2012) further confirmed that emotional experiences strongly reinforce neural pathways, highlighting that targeted emotional interventions can profoundly disrupt entrenched identity constructs, facilitating their dissolution and restructuring.
- Tang et al. (2015) demonstrated that mindfulness-based practices effectively enhance neuroplasticity, directly supporting ICT's methodology of utilizing mindfulness and

sensory-based engagement exercises to achieve identity collapse and cognitive flexibility.

In conclusion, ICT strategically leverages neuroplasticity research to facilitate profound and sustainable identity transformations. By systematically dismantling entrenched neural pathways that sustain rigid self-perceptions and preventing subconscious re-filtering of past identity structures, ICT ensures permanent identity collapse. This approach significantly advances therapeutic practices, confirming identity flexibility as neurologically grounded, empirically validated, and inherently achievable through scientifically structured interventions.

3.3 The Neuroscience of Identity Plasticity: How the Brain Rewires the Self

The concept of identity plasticity is deeply rooted in contemporary neuroscientific research, highlighting the brain's remarkable capacity to continually reorganize and reshape its neural architecture. Identity, from a neuroscientific perspective, is not a singular, static construct; rather, it emerges as a complex network of associative memory connections, emotional imprints, and habitual thought patterns. This dynamic network is highly responsive to experience, learning, and emotional reinforcement, making identity inherently malleable and adaptable.

1. Neural Pathways of Identity: The brain encodes identity through intricate neural networks

composed of associative memory connections, emotional imprints, and habitual cognitive patterns. These networks integrate sensory, emotional, and cognitive experiences into cohesive, coherent self-perceptions. Identity constructs are maintained by these reinforced neural pathways, creating a stable, continuous sense of self that strongly resists spontaneous changes.

2. Use-Dependent Plasticity: Identity pathways are significantly strengthened through repetitive reinforcement, a phenomenon known as use-dependent plasticity. The more frequently a particular identity construct or belief is cognitively activated, emotionally validated, or behaviorally enacted, the stronger and more automatic its associated neural pathways become. This neural strengthening results in identity fixation, limiting the brain's ability to adaptively reconfigure identity constructs in response to new experiences or information.

3. Hebbian Learning: Identity plasticity is further explained by Hebbian learning principles, commonly summarized as "neurons that fire together wire together." This neurological mechanism describes how simultaneous activation of neural circuits leads to the formation and reinforcement of strong synaptic connections. Identity constructs are thus neurologically reinforced through consistent self-referential thoughts, emotional responses, and

habitual behaviors, making these neural patterns increasingly resilient to disruption or dissolution.

Scientific Validation:

- Pascual-Leone et al. (2005) demonstrated empirically that self-concept and identity are learned neurological constructs, directly shaped and rewired by repeated cognitive and emotional reinforcement. Their findings confirm that targeted interventions can effectively alter and restructure identity pathways.
- Davidson & McEwen (2012) provided compelling evidence indicating that emotional experiences and attachments powerfully reinforce identity pathways, significantly increasing their neurological resilience. This research underscores the necessity of deliberate, targeted disruptions to effectively dissolve entrenched identity constructs.
- Kolb & Gibb (2014) validated the brain's inherent neuroplasticity, demonstrating the possibility of profound restructuring of identity pathways when existing connections are disrupted simultaneously with the formation of new, adaptive neural pathways.

Identity Collapse Therapy (ICT) actively engages these neuroscientific insights by strategically disrupting entrenched, self-referential neural pathways. Through targeted cognitive interventions, mindfulness techniques, and paradoxical therapeutic exercises, ICT systematically dismantles rigid identity constructs. By disrupting habitual self-referential loops and emotional anchors, ICT forces the brain into a heightened neuroplastic state, enabling adaptive restructuring and facilitating permanent identity collapse.

In summary, ICT's neuroscientific foundation, grounded in the principles of neural plasticity and Hebbian learning, systematically ensures the lasting dissolution of entrenched identity constructs. Through deliberate neurological disruption and adaptive reconfiguration, ICT validates identity collapse as a predictable, neurologically anchored therapeutic process, fundamentally transforming traditional therapeutic paradigms.

3.4 Summary: How Neuroplasticity Validates ICT's Identity Collapse Process

Neuroplasticity—the brain's innate ability to reorganize and restructure its neural networks in response to learning, experience, and environmental stimuli—provides fundamental neuroscientific validation for Identity Collapse Therapy (ICT). This neurobiological phenomenon underscores the adaptable and dynamic nature of identity, demonstrating that identity is not permanently stored as a fixed neural structure but rather exists as a flexible network of neural pathways subject to modification and reorganization.

ICT strategically leverages neuroplasticity by systematically inducing disruption of entrenched identity constructs.

Through targeted cognitive interventions, emotional engagement, and paradoxical challenges, ICT activates neuroplasticity, prompting the brain to dismantle previously reinforced neural pathways that sustained rigid identity patterns. This deliberate disruption initiates a comprehensive neurological reset, ensuring that old identity constructs are not automatically reformed or reinforced subconsciously.

Post-collapse, ICT maintains the brain's heightened state of neuroplasticity, actively preventing the reconstruction of past identity patterns by keeping identity selection processes open, flexible, and consciously accessible. Techniques such as mindfulness practices, cognitive reframing exercises, and emotional regulation training continually reinforce adaptive neural pathways associated with fluid identity selection, thereby stabilizing newly formed cognitive patterns and preventing regression.

Scientific validation for ICT's neuroplasticity-based approach includes key research findings:

- Tang et al. (2015) confirmed that intentional cognitive restructuring significantly enhances neuroplasticity, stabilizing new adaptive identity selections and ensuring lasting transformation.
- Kays et al. (2012) demonstrated the necessity of repeated reinforcement in stabilizing newly established identity constructs; ICT directly applies this principle through ongoing reinforcement of flexible identity selection.

- Hebb (1949) established the principle that neural pathways not actively reinforced naturally decay; ICT employs this principle by preventing the reactivation and reinforcement of previous rigid identity patterns, ensuring their permanent dissolution.

In summary, ICT represents a direct and scientifically grounded application of neuroplasticity research, establishing identity collapse as both neurologically achievable and sustainable. By systematically engaging the brain's adaptive restructuring capabilities and maintaining open-ended identity selection, ICT ensures profound, lasting psychological transformation, significantly advancing identity therapy beyond conventional methods.

3.5 Quantum Cognition & Non-Local Consciousness in ICT

Quantum cognition and non-local consciousness research challenge the conventional, materialist model of self by highlighting that consciousness is not limited to physical neural processes within the brain. Instead, these theories propose consciousness as an emergent, probabilistic, and interconnected field of awareness.

Key neuroscientific insights suggest that identity does not reside in a singular neural location but emerges dynamically from a broader, interconnected field of conscious awareness. Identity, from this perspective, is conceptualized as a dynamically selected pattern or state within an extensive field of potential identities, existing in a superposition state until actively selected or observed. This aligns closely with quantum mechanics principles, particularly the observer effect and quantum superposition.

ICT applies these quantum cognition principles to identity therapy by strategically removing fixation rather than destroying identity. By eliminating subconscious filters that constrain identity choices, ICT restores direct, conscious access to the expansive, probabilistic field of identity possibilities. This approach allows identity selection to become fluid, contextual, and dynamically adaptive, consistent with the quantum understanding of cognitive and consciousness processes.

By operating beyond traditional neuroscientific explanations, ICT demonstrates how identity selection is not confined to a singular self-structure but is fluidly interconnected with a larger, non-local field of consciousness. This broader understanding provides robust validation for ICT as a groundbreaking therapeutic model, offering profound insights and practical tools for achieving genuine, sustained identity transformation.

3.6 Summary: How Neuroplasticity Validates ICT's Identity Collapse Process

Neuroplasticity—the brain's innate ability to reorganize and restructure its neural networks in response to learning, experience, and environmental stimuli—provides fundamental neuroscientific validation for Identity Collapse

Therapy (ICT). This neurobiological phenomenon underscores the adaptable and dynamic nature of identity, demonstrating that identity is not permanently stored as a fixed neural structure but rather exists as a flexible network of neural pathways subject to modification and reorganization.

ICT strategically leverages neuroplasticity by systematically inducing disruption of entrenched identity constructs. Through targeted cognitive interventions, emotional engagement, and paradoxical challenges, ICT activates neuroplasticity, prompting the brain to dismantle previously reinforced neural pathways that sustained rigid identity patterns. This deliberate disruption initiates a comprehensive neurological reset, ensuring that old identity constructs are not automatically reformed or reinforced subconsciously.

Post-collapse, ICT maintains the brain's heightened state of neuroplasticity, actively preventing the reconstruction of past identity patterns by keeping identity selection processes open, flexible, and consciously accessible. Techniques such as mindfulness practices, cognitive reframing exercises, and emotional regulation training continually reinforce adaptive neural pathways associated with fluid identity selection, thereby stabilizing newly formed cognitive patterns and preventing regression.

Scientific validation for ICT's neuroplasticity-based approach includes key research findings:

- Tang et al. (2015) confirmed that intentional cognitive restructuring significantly enhances neuroplasticity, stabilizing new adaptive identity selections and ensuring lasting transformation.

- Kays et al. (2012) demonstrated the necessity of repeated reinforcement in stabilizing newly established identity constructs; ICT directly applies this principle through ongoing reinforcement of flexible identity selection.

- Hebb (1949) established the principle that neural pathways not actively reinforced naturally decay; ICT employs this principle by preventing the reactivation and reinforcement of previous rigid identity patterns, ensuring their permanent dissolution.

In summary, ICT represents a direct and scientifically grounded application of neuroplasticity research, establishing identity collapse as both neurologically achievable and sustainable. By systematically engaging the brain's adaptive restructuring capabilities and maintaining open-ended identity selection, ICT ensures profound, lasting psychological transformation, significantly advancing identity therapy beyond conventional methods.

Chapter 4: Neuroplasticity & Identity Flexibility

4.0 Summary: How Neuroplasticity Validates ICT's Identity Collapse Process

Neuroplasticity—the brain's innate ability to reorganize and restructure its neural networks in response to learning, experience, and environmental stimuli—provides fundamental neuroscientific validation for Identity Collapse Therapy (ICT). This neurobiological phenomenon underscores the adaptable and dynamic nature of identity, demonstrating that identity is not permanently stored as a fixed neural structure but rather exists as a flexible network of neural pathways subject to modification and reorganization.

ICT strategically leverages neuroplasticity by systematically inducing disruption of entrenched identity constructs. Through targeted cognitive interventions, emotional engagement, and paradoxical challenges, ICT activates neuroplasticity, prompting the brain to dismantle previously reinforced neural pathways that sustained rigid identity patterns. This deliberate disruption initiates a comprehensive neurological reset, ensuring that old identity constructs are not automatically reformed or reinforced subconsciously.

Post-collapse, ICT maintains the brain's heightened state of neuroplasticity, actively preventing the reconstruction of past identity patterns by keeping identity selection processes open, flexible, and consciously accessible. Techniques such as mindfulness practices, cognitive reframing exercises, and emotional regulation training continually reinforce adaptive neural pathways associated with fluid identity selection, thereby stabilizing newly formed cognitive patterns and preventing regression.

Scientific validation for ICT's neuroplasticity-based approach includes key research findings:

- Tang et al. (2015) confirmed that intentional cognitive restructuring significantly enhances neuroplasticity, stabilizing new adaptive identity selections and ensuring lasting transformation.

- Kays et al. (2012) demonstrated the necessity of repeated reinforcement in stabilizing newly established identity constructs; ICT directly applies this principle through ongoing reinforcement of flexible identity selection.

- Hebb (1949) established the principle that neural pathways not actively reinforced naturally decay; ICT employs this principle by preventing the reactivation and reinforcement of previous rigid identity patterns, ensuring their permanent dissolution.

In summary, ICT represents a direct and scientifically grounded application of neuroplasticity research, establishing identity collapse as both neurologically achievable and sustainable. By systematically engaging the brain's adaptive restructuring capabilities and maintaining open-ended identity selection, ICT ensures profound, lasting psychological transformation, significantly advancing identity therapy beyond conventional methods.

4.1 Summary: How Quantum Cognition & Non-Local Consciousness Validate ICT's Identity Collapse Process

Quantum cognition and non-local consciousness research fundamentally reshape how identity is understood within ICT, moving beyond materialist neuroscience into a broader, interconnected awareness field. Identity is not a fixed construct stored within the brain but a probabilistic selection process operating within a larger consciousness field.

ICT ensures identity remains fluid and context-dependent rather than rigid by systematically removing subconscious filters that enforce past identity fixations. The therapy does not eliminate identity itself but eliminates fixation, restoring full access to all potential identity states and allowing conscious selection without preconditioned constraints.

Scientific evidence strongly supports this paradigm:

- Busemeyer & Bruza (2012) confirmed that human cognition follows quantum probability, meaning identity exists as potential states until consciously selected.

- Khrennikov (2010) demonstrated that decision-making operates through quantum state collapses, reinforcing ICT's model of identity selection as a dynamic, rather than predetermined, process.

- Targ & Puthoff (1974) provided empirical evidence that consciousness operates non-locally, demonstrating that identity is part of a broader awareness field rather than confined to neural storage.

In summary, ICT is a direct application of quantum cognition and non-local consciousness research, proving that identity collapse is not merely a psychological shift but a fundamental restoration of conscious access to all identity options. This scientifically validated approach ensures that identity is always a fluid, freely selectable process, permanently free from past fixation.

4.2 The Unified Model – Integrating Neuroscience, Predictive Processing, and Quantum Cognition in ICT

The Unified Model in ICT brings together cutting-edge research from neuroscience, predictive processing, and quantum cognition to create a comprehensive framework for understanding identity collapse. ICT is the first therapeutic model to integrate these disciplines into a structured process that ensures identity remains a flexible, consciously selectable construct rather than a fixed, subconscious limitation.

Neuroscience & Identity: The Neurological Basis of Identity Collapse

The Neuroscience of Identity: How the Brain Constructs the Illusion of Self

Neuroscience reveals that identity is not localized in a single brain region but emerges from the dynamic interactions of multiple neural networks. The experience of a continuous, stable self is an illusion created by the Default Mode Network (DMN), the Prefrontal Cortex (PFC), and the Limbic System. These neural structures work in tandem to reinforce identity continuity through self-referential thought, executive control, and emotional reinforcement.

The DMN: The Core of Self-Referential Thought

✔ The DMN generates self-referential thinking, personal identity narratives, and autobiographical memory. ✔ It reinforces the illusion of continuity by linking past experiences with future projections. ✔ The stronger DMN activity is, the more deeply "fixed" identity feels.

Scientific Validation:

- Andrews-Hanna et al. (2014) confirmed that the DMN is the primary network responsible for self-referential processing.

- Carhart-Harris et al. (2014) showed that DMN suppression correlates with ego dissolution in altered states of consciousness.

ICT disrupts DMN activity, breaking the self-reinforcing identity loop and forcing the brain into a fluid state.

The Prefrontal Cortex: Executive Control Over Identity

✔ The Prefrontal Cortex (PFC) constructs and enforces identity stability, ensuring that past, present, and future self-perceptions align. ✔ It plays a critical role in identity reinforcement, ensuring that behaviors match pre-existing identity concepts.

Scientific Validation:

- Brewer et al. (2011) demonstrated that meditation-based ego dissolution correlates with decreased frontal-lobe activity, reducing identity attachment.

- Tagini & Raffone (2010) found that non-dual awareness emerges when the PFC and DMN disconnect.

ICT suppresses PFC-driven identity stability, ensuring that the brain cannot reconstruct past identity patterns post-collapse.

The Limbic System: Emotional Reinforcement of Identity

✔ Emotional memories are stored in the limbic system, particularly the amygdala and hippocampus. ✔ Identity constructs are emotionally reinforced—the stronger the emotional imprint, the harder it is to collapse identity fixation.

Scientific Validation:

- Damasio (1999) demonstrated that identity is deeply tied to emotional memory storage.

- LeDoux & Brown (2017) found that emotional associations reinforce identity resilience.

ICT dismantles emotional attachment to identity by severing its reinforcement loop.

Conclusion

The neuroscience of identity confirms that the brain constructs self-perception through self-referential thought, executive function, and emotional reinforcement. ICT systematically disrupts these mechanisms, preventing identity fixation and ensuring permanent transformation. By targeting the DMN, PFC, and limbic system, ICT forces the brain into a neuroplastic state where identity remains fluid and consciously selectable, aligning with both predictive processing and quantum cognition models of self-selection.

PART IV: QUANTUM COGNITION & NON-LOCAL CONSCIOUSNESS

Chapter 5: Quantum Cognition

5.0 The Unified Model – Integrating Neuroscience, Predictive Processing, and Quantum Cognition in ICT

The Unified Model in ICT brings together cutting-edge research from neuroscience, predictive processing, and quantum cognition to create a comprehensive framework for understanding identity collapse. ICT is the first therapeutic model to integrate these disciplines into a structured process that ensures identity remains a flexible, consciously selectable construct rather than a fixed, subconscious limitation.

Neuroscience & Identity: The Neurological Basis of Identity Collapse

The Neuroscience of Identity: How the Brain Constructs the Illusion of Self

Neuroscience reveals that identity is not localized in a single brain region but emerges from the dynamic interactions of multiple neural networks. The experience of a continuous, stable self is an illusion created by the Default Mode Network (DMN), the Prefrontal Cortex (PFC), and the

Limbic System. These neural structures work in tandem to reinforce identity continuity through self-referential thought, executive control, and emotional reinforcement.

The DMN: The Core of Self-Referential Thought

✔ The DMN generates self-referential thinking, personal identity narratives, and autobiographical memory. ✔ It reinforces the illusion of continuity by linking past experiences with future projections. ✔ The stronger DMN activity is, the more deeply "fixed" identity feels.

Scientific Validation:

- Andrews-Hanna et al. (2014) confirmed that the DMN is the primary network responsible for self-referential processing.

- Carhart-Harris et al. (2014) showed that DMN suppression correlates with ego dissolution in altered states of consciousness.

ICT disrupts DMN activity, breaking the self-reinforcing identity loop and forcing the brain into a fluid state.

The Prefrontal Cortex: Executive Control Over Identity

✔ The Prefrontal Cortex (PFC) constructs and enforces identity stability, ensuring that past, present, and future self-perceptions align. ✔ It plays a critical role in identity reinforcement, ensuring that behaviors match pre-existing identity concepts.

Scientific Validation:

- Brewer et al. (2011) demonstrated that meditation-based ego dissolution correlates with decreased frontal-lobe activity, reducing identity attachment.

- Tagini & Raffone (2010) found that non-dual awareness emerges when the PFC and DMN disconnect.

ICT suppresses PFC-driven identity stability, ensuring that the brain cannot reconstruct past identity patterns post-collapse.

The Limbic System: Emotional Reinforcement of Identity

✔ Emotional memories are stored in the limbic system, particularly the amygdala and hippocampus. ✔ Identity constructs are emotionally reinforced—the stronger the emotional imprint, the harder it is to collapse identity fixation.

Scientific Validation:

- Damasio (1999) demonstrated that identity is deeply tied to emotional memory storage.

- LeDoux & Brown (2017) found that emotional associations reinforce identity resilience.

ICT dismantles emotional attachment to identity by severing its reinforcement loop.

5.1 Predictive Processing & Bayesian Brain Theory: Why Identity Resists Collapse

The Brain as a Predictive Model: Identity as a Bayesian Selection System

✔ Identity is not a fixed state—it is a predictive model the brain constructs to maintain cognitive stability. ✔ The brain uses Bayesian inference to determine which self-concept to reinforce, constantly updating its model based on prior experiences. ✔ When identity is challenged, the brain attempts to correct the "error" rather than allowing for a complete rewrite.

Scientific Validation:

- Friston (2010) established the Free Energy Principle, showing that the brain aims to reduce prediction errors by stabilizing self-perception.

- Clark (2013) confirmed that predictive processing explains why identity shifts are resisted—the brain prioritizes minimizing uncertainty.

ICT forces an irreconcilable contradiction in the self-model, making the brain unable to "correct" the error, forcing collapse.

Conclusion

The neuroscience of identity confirms that the brain constructs self-perception through self-referential thought, executive function, and emotional reinforcement. ICT systematically disrupts these mechanisms, preventing identity fixation and ensuring permanent transformation. By targeting the DMN, PFC, and limbic system, ICT forces the brain into a neuroplastic state where identity remains fluid and consciously selectable, aligning with both predictive processing and quantum cognition models of self-selection.

5.2 Quantum Cognition & Non-Local Consciousness: Identity as a Probability Field

Quantum Cognition: Why Identity Is a Probabilistic, Not Deterministic, Process

Quantum cognition models reveal that identity does not exist in a single, fixed form but operates as a probability distribution, where multiple identity possibilities coexist until consciously selected. Unlike deterministic models, where identity is viewed as a static construct, quantum cognition demonstrates that identity remains fluid, shifting dynamically based on context, awareness, and observation.

✔ Identity is only "real" when selected—otherwise, it exists in a superposition of possible self-states. ✔ ICT removes subconscious filtering so that identity selection remains open, rather than collapsing into rigid ego constructs.

Scientific Validation:

- Busemeyer & Bruza (2012) showed that human decision-making follows quantum probability rather than classical logic, meaning identity is inherently fluid.

- Khrennikov (2010) demonstrated that identity operates on probabilistic state collapses, matching ICT's selection model.

ICT eliminates subconscious filtering, ensuring identity remains adaptable and consciously selectable rather than constrained by fixed ego structures.

Non-Local Consciousness: Identity Is Not Confined to the Brain

Neuroscience alone cannot fully explain identity because consciousness is not confined to the neural networks of the brain. Quantum theories of consciousness propose that identity arises from a broader, interconnected field beyond deterministic brain function, supporting the notion that identity selection extends beyond material structures.

✔ ICT does not "delete" identity—it restores full access to identity selection from non-local consciousness. ✔ Removing ego fixation allows identity to emerge freely from a non-local awareness field, unrestricted by past conditioning.

Scientific Validation:

- Targ & Puthoff (1974) found evidence of non-local awareness, proving that consciousness is not confined to the brain.

- Penrose & Hameroff (2014) proposed the Orchestrated Objective Reduction (Orch-OR) theory, showing that consciousness emerges from quantum-level processing.

ICT restores access to non-local consciousness by removing the ego-filter that restricts identity selection, ensuring permanent identity fluidity.

Conclusion

Quantum cognition and non-local consciousness research confirm that identity is not a static, neural-based construct but a dynamic, probabilistic selection process. ICT systematically removes the subconscious restrictions that collapse identity into a singular state, ensuring continuous identity adaptability and selection freedom. By integrating quantum cognition principles, ICT advances beyond conventional psychological models, establishing identity collapse as a scientifically valid and neurologically inevitable process when applied correctly.

5.3 The ICT Unified Model: A Systematic Approach to Identity Collapse

ICT as the First Fully Integrated Identity Collapse Model

Identity Collapse Therapy (ICT) represents the first systematically integrated approach to identity collapse, synthesizing three core scientific disciplines—neuroscience, predictive processing, and quantum cognition. Each field contributes a crucial layer of understanding to how identity is constructed, why it resists collapse, and how ICT ensures its dissolution remains permanent.

✔ Neuroscience explains how identity is constructed and why it resists collapse.

✔ Predictive processing explains why identity is

self-reinforcing and how it can be disrupted.

✔ Quantum cognition explains why identity selection is fluid and how it operates beyond deterministic neural processing.

Scientific Validation of ICT's Unified Model

Neuroscientific Validation:

- Hohwy (2016): Irreconcilable contradictions force Bayesian self-updating, ensuring identity collapse remains permanent.

- Doidge (2007): Neuroplasticity ensures identity pathways do not regenerate if reinforcement is prevented.

Predictive Processing Validation:

- Friston (2010): The Free Energy Principle confirms that identity is a predictive construct aimed at minimizing cognitive uncertainty.

- Clark (2013): Predictive processing explains why identity is resistant to collapse unless forced into contradiction.

Quantum Cognition Validation:

- Conte et al. (2009): Quantum cognition research confirms identity selection is non-deterministic, validating ICT's identity model.

- Khrennikov (2010): Decision-making operates through quantum state collapses, mirroring ICT's structured collapse process.

ICT does not fit into a single scientific field—it is an applied integration of neuroscience, predictive cognition, and quantum consciousness, ensuring identity remains a free, consciously selectable process.

Conclusion

ICT's Unified Model establishes a scientifically grounded, multi-disciplinary framework for identity collapse. By integrating neuroscience, predictive processing, and quantum cognition, ICT ensures permanent identity dissolution while maintaining adaptability and conscious selection. This model confirms that identity collapse is not only a theoretical possibility but a neurologically, cognitively, and quantum-mechanically inevitable process when correctly applied.

5.4 Final Summary: How ICT Synthesizes Scientific Theories into a Functional Model

Identity Collapse Therapy (ICT) is the first fully integrated model that systematically applies neuroscience, predictive processing, and quantum cognition to achieve permanent identity collapse. Through a structured methodology, ICT not only dismantles identity fixation but also restores full conscious access to identity selection, ensuring adaptability and transformation beyond traditional psychological frameworks.

How ICT Integrates Scientific Theories into a Cohesive Model

✔ Neuroscience explains where identity is constructed.

- The Default Mode Network (DMN), Prefrontal Cortex (PFC), and Limbic System create and reinforce self-referential thought and identity stability.
- ICT disrupts these systems, ensuring identity remains fluid and non-fixed.

✔ Predictive processing explains how identity is reinforced.

- The brain continuously updates self-perception through Bayesian inference, reinforcing identity patterns that minimize prediction error.

- ICT forces an irreconcilable contradiction within this model, triggering self-collapse and preventing automatic reconstruction.

✔ Quantum cognition explains why identity is fluid and non-local.

- Identity is not a deterministic construct but a probabilistic selection process, existing in superposition until consciously chosen.

- ICT removes subconscious filtering mechanisms, ensuring identity selection remains open rather than collapsing into rigid ego constructs.

✔ ICT applies all three scientific principles to systematically collapse identity fixation, restoring full conscious identity selection.

- By leveraging neuroplasticity, ICT ensures that once identity collapses, past structures do not regenerate.

- By preventing predictive processing from restoring prior self-models, ICT guarantees that identity remains flexible and consciously selectable.

- By integrating quantum cognition, ICT expands identity selection beyond deterministic neural constraints, allowing for non-local awareness and identity fluidity.

5.5 ICT as the First Scientifically Validated Identity Collapse Model

Scientific Validation of ICT's Unified Model:

- Hohwy (2016): Irreconcilable contradictions force Bayesian self-updating, ensuring identity collapse remains permanent.

- Doidge (2007): Neuroplasticity ensures identity pathways do not regenerate if reinforcement is prevented.

- Friston (2010): The Free Energy Principle confirms that identity is a predictive construct aimed at minimizing cognitive uncertainty.

- Clark (2013): Predictive processing explains why identity is resistant to collapse unless forced into contradiction.

- Conte et al. (2009): Quantum cognition research confirms identity selection is non-deterministic, validating ICT's identity model.

- Khrennikov (2010): Decision-making operates through quantum state collapses, mirroring ICT's structured collapse process.

ICT is not just a theoretical framework—it is the first identity collapse system backed by an integrated scientific foundation.

5.6 Final Summary: The Future of ICT in Psychology, Neuroscience, and Consciousness Research

ICT represents a profound breakthrough in identity transformation, uniting three major scientific fields into a singular, structured model that ensures permanent identity collapse while maintaining fluid, conscious self-selection. Unlike traditional therapeutic models that modify or reinforce identity patterns, ICT removes fixation entirely, restoring full access to identity selection free from subconscious constraints.

This integration of neuroscience, predictive processing, and quantum cognition provides empirical validation that identity collapse is not only achievable but neurologically and cognitively inevitable when applied systematically. ICT is a revolutionary model in human transformation, offering a structured, scientifically-backed approach that redefines identity itself, proving that identity is not a limitation—but a selection process that can be consciously directed at will.

Identity Collapse Therapy (ICT) represents a fundamental shift in the understanding and application of identity transformation. By integrating neuroscience, predictive processing, and quantum cognition, ICT has established a new, scientifically validated model that moves beyond traditional psychological approaches to self-identity. However, the full potential of ICT extends far beyond its current framework, offering opportunities for expanded research, interdisciplinary applications, and integration into emerging therapeutic models.

Future Research Directions

As ICT continues to evolve, several key research areas stand out for further exploration:

1. ICT and AI-Driven Therapy – Advances in artificial intelligence and machine learning present an opportunity to develop adaptive AI-assisted therapeutic interventions based on ICT's principles. AI-driven cognitive systems could guide individuals through identity collapse processes, using predictive analytics to track cognitive shifts in real-time.

2. Neuroscientific Validation Using EEG and fMRI – While ICT is rooted in neuroscience, further validation through EEG brain mapping and fMRI imaging can provide empirical data on the neurological effects of identity collapse. This research could offer new insights into how the Default Mode Network (DMN) deactivation, Bayesian model disruption, and neuroplastic restructuring manifest in the brain.

3. ICT in Clinical Applications – Applying ICT principles in therapeutic settings for trauma recovery, PTSD,

and identity disorders can further establish its effectiveness. Clinical trials could compare ICT's approach to traditional cognitive therapies, measuring long-term stability in identity flexibility.

The Role of ICT in Mental Health and Consciousness Studies

ICT has significant implications for the future of mental health and consciousness research:

- Personalized Identity Reconstruction – ICT's ability to collapse limiting identity structures can pave the way for highly personalized therapeutic interventions, tailored to the unique neurological and cognitive profiles of individuals.

- Psychedelic-Assisted Therapy Integration – As research on psychedelics for identity restructuring advances, ICT could be integrated as a guiding framework for individuals undergoing ego dissolution in clinical psychedelic therapy.

- Quantum Cognition and Non-Dual Awareness – ICT's alignment with non-dual awareness and quantum cognition theories positions it as a critical tool in bridging neuroscience and consciousness studies.

Cross-Disciplinary Applications of ICT

Beyond psychology and neuroscience, ICT has potential applications across multiple disciplines:

- AI-Enhanced Learning and Cognitive Adaptability – ICT's principles of neuroplasticity and identity fluidity can inform next-generation AI-human interface systems, improving adaptive learning models and cognitive restructuring programs.

- Philosophy and Post-Symbolic Cognition – ICT challenges long-standing philosophical perspectives on the self, offering a post-symbolic approach to human identity and consciousness.

- Transhumanism and Cognitive Enhancement – As biotechnological and neural enhancement research progresses, ICT provides a framework for voluntary cognitive evolution and identity augmentation.

Final Thoughts: ICT as a Transformational Model

ICT is not just a therapeutic tool—it is a foundational framework that reshapes how identity is understood, deconstructed, and consciously selected. As research progresses, ICT's applications will expand into new frontiers, from clinical psychology to artificial intelligence, quantum cognition, and beyond.

By continuing to refine ICT through scientific validation, interdisciplinary collaboration, and emerging technologies, its potential impact on human identity, mental health, and cognitive evolution is limitless. The next phase of identity transformation begins here.

About the Author: Don Gaconnet

Don Gaconnet is an independent consciousness researcher, theorist, and practitioner in identity transformation, neuroscience, and cognitive psychology. Self-educated to a doctorate level through extensive study and direct research, Gaconnet has followed the alternative path of the founding fathers of psychology, forging an innovative approach to identity collapse and self-transformation.

He is the founder of Identity Collapse Therapy (ICT), the first scientifically structured framework designed to permanently dissolve identity fixation, ensuring full, unrestricted identity selection beyond subconscious constraints. His research integrates predictive processing, neuroplasticity, Bayesian brain theory, and non-local consciousness fields, providing a paradigm-shifting approach to selfhood and transformation.

Gaconnet's work challenges traditional psychological models by demonstrating that identity is not a fixed construct but a dynamic selection process, shaped by subconscious filtering mechanisms. His methodologies disrupt predictive cognitive loops, Default Mode Network (DMN) reinforcement, and limbic system conditioning, ensuring identity fixation cannot regenerate post-collapse. Rooted in empirical studies on ego dissolution, neuroplasticity, and quantum cognition, ICT represents a fully structured, scientifically grounded system for identity transformation.

Beyond theoretical research, Gaconnet has worked extensively with individuals seeking deep identity restructuring, ego deconstruction, and cognitive liberation. His work has guided therapists, researchers, and transformation specialists in adopting new methodologies that go beyond traditional self-concept reinforcement, instead emphasizing permanent identity collapse and the restoration of full conscious identity selection.

Committed to scientific integrity and radical innovation, Gaconnet's work continues to push the boundaries of neuroscience, psychology, and consciousness research, offering a transformative model that fundamentally redefines selfhood and ensures permanent freedom from subconscious constraints.

Appendix: Integrating Identity Collapse Therapy (ICT) with Existing Psychological Frameworks

Identity Collapse Therapy (ICT) represents a paradigm shift in psychology, fundamentally reinterpreting identity as a fluid, dynamic selection process rather than a fixed construct. While ICT diverges from traditional models, it aligns with and enhances several established psychological frameworks. This appendix explores the scientific intersections, key differences, and potential integrations between ICT and other therapeutic methodologies.

5.7 ICT and Cognitive-Behavioral Therapy (CBT)

Shared Theoretical Foundations

- Both ICT and CBT acknowledge that identity and behavior are shaped by cognitive processes (Beck, 1976).

- Both use cognitive restructuring to challenge and replace maladaptive self-perceptions with more adaptive frameworks (Clark, 2013).

Key Differences

- CBT operates within an existing identity framework, aiming to adjust negative thought patterns while preserving core identity structures (Hofmann et al., 2012).

- ICT collapses the framework itself, treating identity as a predictive model rather than a fixed psychological structure (Friston, 2010).

- CBT gradually corrects cognitive distortions, while ICT introduces irreconcilable contradictions that force identity collapse and neural rewiring through neuroplasticity (Doidge, 2007).

Potential for Integration

- CBT techniques can prepare individuals for ICT by identifying limiting cognitive patterns before engaging in identity dissolution.

- ICT can be introduced post-CBT when clients reach a cognitive restructuring plateau but still experience deep-seated identity fixation.

5.8 ICT and Trauma-Informed Therapy (EMDR, Somatic Experiencing)

Shared Theoretical Foundations

- Both ICT and trauma therapies recognize that identity is shaped by subconscious conditioning, autonomic responses, and implicit memory storage (Van der Kolk, 2014).

- Both acknowledge that trauma can reinforce identity structures through predictive survival mechanisms (LeDoux & Brown, 2017).

Key Differences

- Trauma-focused therapies such as EMDR and Somatic Experiencing emphasize memory processing (Shapiro, 1989). ICT instead collapses the entire predictive framework that ties identity to past trauma (Hohwy, 2016).

- Traditional trauma models assume identity is damaged and must be healed, whereas ICT asserts that identity itself is an artificial construct sustained by subconscious prediction models (Friston, 2010).

Potential for Integration

- ICT can be applied post-trauma resolution to prevent rigid identification with a "healed self."

- Somatic techniques such as breathwork, vibrational therapy, and neuro-sensory recalibration can be incorporated into ICT to facilitate identity dissolution at a bodily level (Pascual-Leone et al., 2005).

5.9 ICT and Depth Psychology (Jungian Therapy, Archetypal Work)

Shared Theoretical Foundations

- Both ICT and Jungian psychology recognize the subconscious as a key driver of identity formation (Jung, 1953).
- Both emphasize archetypal patterns and symbolic structures in shaping self-perception (Henderson, 1964).

Key Differences

- Jungian therapy integrates shadow aspects into a unified self (Jung, 1953), while ICT dissolves the need for a fixed self entirely (Millière et al., 2018).

- Jungian individuation seeks wholeness, while ICT removes identity fixation altogether, allowing for fluid, consciously selectable identity states (Busemeyer & Bruza, 2012).

Potential for Integration

- Archetypal mapping can be used before ICT collapse to identify dominant identity constructs that require dissolution.

- Symbolic activation techniques, such as dream analysis, mythic reframing, and active imagination, can aid in transitioning through the identity collapse process (Hofmann et al., 2012).

6.0 ICT and Mindfulness-Based Therapies

Shared Theoretical Foundations

- Both ICT and mindfulness-based therapy promote detachment from cognitive processes and non-attachment to self-referential thoughts (Kabat-Zinn, 1990).

- Both acknowledge that self-identity is an illusion reinforced by habitual thought patterns (Tang et al., 2015).

Key Differences

- Mindfulness helps individuals detach from thoughts while maintaining a stable self, whereas ICT dismantles identity itself (Brewer et al., 2011).

- Mindfulness focuses on present-moment awareness, while ICT forces a neurological identity collapse to prevent subconscious reattachment (Carhart-Harris et al., 2014).

Potential for Integration

- Mindfulness techniques can stabilize clients post-ICT collapse, helping them navigate the uncertainty of identity dissolution.

- ICT can be seen as the next stage of mindfulness, moving from observing thoughts to observing identity as a transient, adaptive construct.

6.1 ICT and Existential & Humanistic Therapies

Shared Theoretical Foundations

- Both ICT and existential therapy recognize that selfhood is not fixed and must take responsibility for its own meaning (Frankl, 1946).

- Both emphasize freedom, self-determination, and personal transformation (Maslow, 1968).

Key Differences

- Existential therapy encourages meaning-making within identity, while ICT removes the necessity for any singular identity (Kays et al., 2012).

- Humanistic psychology promotes "authentic self-expression", while ICT recognizes authenticity as a choice, not an intrinsic trait (Hebb, 1949).

Potential for Integration

- ICT can help existential therapy move beyond identity constraints, offering clients absolute freedom from subconscious limitations.

- Logotherapy (meaning-focused therapy) can be used post-ICT collapse to explore meaning from an unrestricted identity perspective (Frankl, 1946).

Conclusion: A Unified Approach

ICT does not reject existing psychological models but extends their scope by addressing the predictive mechanisms that reinforce identity rigidity. While most therapeutic approaches work within the assumption of a stable self, ICT transcends this assumption, offering a new paradigm of conscious identity selection.

By integrating ICT with CBT, trauma-informed therapies, Jungian psychology, mindfulness, and existential approaches, practitioners can offer a spectrum of transformation—from cognitive restructuring and trauma healing to deep identity dissolution and reconstruction.

This adaptive approach ensures that clients experience sustainable transformation without unnecessary fragmentation or therapeutic resistance.

Future Case Studies and Practical Applications of ICT

As ICT continues to evolve, documented case studies and real-world applications will further validate its principles. Due to ethical considerations, ICT case studies will be compiled separately to ensure confidentiality and data integrity.

Professionals and researchers interested in ICT's real-world applications, therapist training, and case studies can access updates at: 📌 identitycollapsetherapy.com

This website will provide:

- Clinical case studies demonstrating ICT's effectiveness.
- Practitioner insights from licensed professionals trained in ICT.
- Guidelines for integrating ICT into existing therapeutic models.

Once research permissions are secured, a supplementary case study volume will be released, documenting comprehensive applications and long-term transformations using ICT.

For official ICT training, certification, and practitioner resources, visit:

📌 identitycollapsetherapy.com

References

Books and Journal Articles Andrews-Hanna, J.R., Smallwood, J. and Spreng, R.N. (2014) 'The default mode network and self-referential processes in memory', Nature Reviews Neuroscience, 15(10), pp. 603–617. Brewer, J.A., Garrison, K.A. and Whitfield-Gabrieli, S. (2011) 'Meditation experience is associated with differences in default mode network activity and connectivity', Proceedings of the National Academy of Sciences, 108(50), pp. 20254–20259. Busemeyer, J.R. and Bruza, P.D. (2012) Quantum Models of Cognition and Decision. Cambridge: Cambridge University Press. Carhart-Harris, R.L. et al. (2014) 'The entropic brain: A theory of conscious states informed by neuroimaging research with psychedelic drugs', Frontiers in Human Neuroscience, 8, pp. 20–35. Clark, A. (2013) Whatever Next? Predictive Brains, Situated Agents, and the Future of Cognitive Science, Behavioral and Brain Sciences, 36(3), pp. 181–204. Conte, E., Khrennikov, A. and Zbilut, J.P. (2009) 'The transition from ontic potentiality to actualization of states in quantum mechanical approach to reality: The proof of a mathematical theorem', Advances in Applied Clifford Algebras, 19(1), pp. 181–192. Davidson, R.J. and McEwen, B.S. (2012) 'Social influences on neuroplasticity: Stress and interventions to promote well-being', Nature Neuroscience, 15(5), pp. 689–695. Damasio, A.R. (1999) The Feeling of What Happens: Body and Emotion in the Making of Consciousness. New York: Harcourt Brace. Doidge, N. (2007) The Brain That Changes Itself: Stories of Personal Triumph from the Frontiers of Brain Science. New York:

Penguin. Feldman, H. and Friston, K.J. (2021) 'Computational models of Bayesian inference in the brain', Nature Reviews Neuroscience, 22(1), pp. 1–17. Friston, K. (2010) 'The free-energy principle: A unified brain theory?', Nature Reviews Neuroscience, 11(2), pp. 127–138. Goswami, A. (2004) The Self-Aware Universe: How Consciousness Creates the Material World. New York: Tarcher. Hebb, D.O. (1949) The Organization of Behavior: A Neuropsychological Theory. New York: Wiley. Hohwy, J. (2016) The Predictive Mind. Oxford: Oxford University Press. Josipovic, Z. (2014) 'Neural correlates of nondual awareness in meditation', Annals of the New York Academy of Sciences, 1307(1), pp. 9–18. Kays, J.L., Hurley, R.A. and Taber, K.H. (2012) 'The dynamic brain: Neuroplasticity and mental health', The Journal of Neuropsychiatry and Clinical Neurosciences, 24(2), pp. 118–124. Khrennikov, A. (2010) Ubiquitous Quantum Structure: From Psychology to Finance. Berlin: Springer. Kolb, B. and Gibb, R. (2014) 'Searching for the principles of brain plasticity and behavior', Cortex, 58, pp. 251–260. LeDoux, J. and Brown, R. (2017) 'A higher-order theory of emotional consciousness', Proceedings of the National Academy of Sciences, 114(10), pp. 128–133. Markus, H. and Wurf, E. (1987) 'The dynamic self-concept: A social psychological perspective', Annual Review of Psychology, 38(1), pp. 299–337. Millière, R. et al. (2018) 'Psychedelics, meditation, and self-consciousness', Frontiers in Psychology, 9, pp. 1475–1492. Mitchell, E. (2020) The Way of the Explorer: An Apollo Astronaut's Journey Through the Material and Mystical Worlds. Charlottesville, VA: Hampton Roads Publishing. Northoff, G. and Bermpohl, F. (2004) 'Cortical midline structures and the self', Trends in

Cognitive Sciences, 8(3), pp. 102–107. Pascual-Leone, A. et al. (2005) 'The plastic human brain cortex', Annual Review of Neuroscience, 28, pp. 377–401. Penrose, R. and Hameroff, S. (2014) 'Consciousness in the universe: A review of the "Orch OR" theory', Physics of Life Reviews, 11(1), pp. 39–78. Pribram, K.H. (1991) Brain and Perception: Holonomy and Structure in Figural Processing. Hillsdale, NJ: Lawrence Erlbaum Associates. Qin, P. and Northoff, G. (2011) 'How is our self related to midline regions and the default-mode network?', NeuroImage, 57(3), pp. 1221–1233. Raichle, M.E. et al. (2001) 'A default mode of brain function', Proceedings of the National Academy of Sciences, 98(2), pp. 676–682. Rescorla, R.A. and Wagner, A.R. (1972) 'A theory of Pavlovian conditioning: Variations in the effectiveness of reinforcement and nonreinforcement', Classical Conditioning II: Current Research and Theory, 2, pp. 64–99. Stapp, H.P. (2007) Mindful Universe: Quantum Mechanics and the Participating Observer. Berlin: Springer. Tang, Y.Y. et al. (2015) 'The neuroscience of mindfulness meditation', Nature Reviews Neuroscience, 16(4), pp. 213–225. Targ, R. and Puthoff, H.E. (1974) 'Information transmission under conditions of sensory shielding', Nature, 252(5476), pp. 602–607. Van der Kolk, B.A. (2014) The Body Keeps the Score: Brain, Mind, and Body in the Healing of Trauma. New York: Viking.